Mortars in
World War II

Mortars in World War II

Artillery for the Infantry

John Norris

Pen & Sword
MILITARY

First published in Great Britain in 2015 by
PEN & SWORD MILITARY
an imprint of
Pen & Sword Books Ltd
47 Church Street
Barnsley
South Yorkshire
S70 2AS

ISBN 978-1-78346-376-3

Typeset by Concept, Huddersfield, West Yorkshire, HD4 5JL.
Printed and bound in England by CPI Group (UK) Ltd, Croydon CR0 4YY.

Pen & Sword Books Ltd incorporates the imprints of Pen & Sword Archaeology,
Atlas, Aviation, Battleground, Discovery, Family History, History, Maritime,
Military, Naval, Politics, Railways, Select, Social History, Transport, True Crime,
and Claymore Press, Frontline Books, Leo Cooper, Praetorian Press,
Remember When, Seaforth Publishing and Wharncliffe.

For a complete list of Pen & Sword titles please contact
PEN & SWORD BOOKS LIMITED
47 Church Street, Barnsley, South Yorkshire, S70 2AS, England
E-mail: enquiries@pen-and-sword.co.uk
Website: www.pen-and-sword.co.uk

DEDICATION

To my wife, Elizabeth, daughter Charlotte and grand-daughters Harriett and Olivia, for their patience and understanding whilst I jotted notes and took measurements of mortars in all weathers.

Contents

Acknowledgements

I would like to extend my sincere thanks to everybody who provided me with reference material, including weapons collectors such as Jeff Wilson who made manuals available to me. My special thanks to John Adams-Graf in America for providing some great images and to Terry Gander for allowing the reproduction of the M19 artwork by the late Lyn Harwood. I am particularly grateful to the Jersey Branch of the Channel Islands Occupation Society for allowing the use of the photograph showing the M19 automatic mortar. I am grateful to the many re-enactors and staff at various regimental museums for allowing me to take photographs. I am most indebted to the staff at the SASC Weapons Collection of the British Army at Warminster in Wiltshire for arranging a visit and allowing images to be reproduced. Finally, conversations and correspondence with fellow weapon enthusiasts such as Andy Colborn and Neil Lawrence, who very kindly provided more information which proved to be most useful.

Introduction

During the Second World War the composite structures of the infantry units making up the armies of the belligerent nations, such as divisions and Corps, altered as the course of fighting changed. The infantrymen serving in these units had to learn how to use a range of weapons from their basic service rifle to anti-tank weapons. In the period of just over twenty years since the end of the First World War, the role of the infantry had changed dramatically with the introduction of a range of new weapons on to the battlefield such as tanks, flamethrowers and anti-tank rifles. Another new weapon which the infantry had to learn to use was the trench mortar, which appeared in a range of calibres. Specialist units were created to operate them.

As the scale of the fighting spread, some armies ceased to exist. The Polish Army became the first victim. When the country surrendered on 27 September 1939 thousands of troops were taken captive and sent off to prisoner of war (PoW) camps. Other victims included France and Belgium, attacked by the German Army in May 1940 using tanks in Panzer divisions which benefitted from the fire support provided by sixty mortars. A year later, when Hitler turned his attention eastwards to the Soviet Union, the Germans had restructured a division and increased the number of mortars to include fifty-four Model 1934 GrW34 mortars of 8.1cm and eighty-four of the smaller leGrW36 5cm calibre mortars as fire support. As the war progressed, other armies such as Britain and America would also make changes by either increasing or decreasing the number of mortars available to each division.

Early on in the war the German Army adopted the practice of taking into service any captured stocks of military equipment considered useful to their war effort. This move was nothing new and had, in fact, been used by Napoleon Bonaparte in the early nineteenth century. Equipment put to use by the German Army in this way included vehicles such as trucks, tanks and armoured cars, along with a range of other weapons and stockpiles of ammunition. This policy applied to the massive piles of weapons and ammunition from the defeated Dutch, Belgium and French armies when they surrendered, and naturally included their stock of mortars. Standard service weapons changed and some were modified during the course of the war. New weapons were also introduced as replacements for the older, obsolete types as they were taken out of service. Mortars, however, because of their simple construction were a range of weapons which remained in service virtually unchanged throughout the war.

Throughout the war, mortars were primarily used by ordinary infantry units, but there were some attempts made to specially modify some types for use by units such as airborne troops who required lightweight weapons. Some of the designs worked. Others were found not to be suitable and never left the drawing board. Mortars were also mounted in armoured vehicles to give mobility to the weapons and this was an area of development led by the German and American armies, both of which produced a series fitted to half-track vehicles. The basic standard infantry mortar was used in combat in all theatres of operations from Europe to the sandy wastes of North Africa, the Russian steppes, with its extremes in weather conditions, and the steaming jungles in the Far East. With no moving parts, it was an ideal weapon for all these conditions.

Mortars are crew-served weapons, which is to say three or four men make up a crew to operate the weapon, with each man serving in a specific role to load or aim the weapon. Some countries such as the Soviet Union developed mortars of very large calibre such as the 160mm M1943 type, but generally most armies used types which were categorised either as 'light', 'medium' or 'heavy'. Larger, highly specialised types were developed such as the *Granatenwerfer* 69 which entered service with the German Army in 1944 and was originally meant to have a calibre of 22cm and known as the sGrW14, but a change in plan reduced the calibre to 21cm. It was late in entering service and only about 200 are believed to have been built. Although categorised as a mortar, its large size and weight of 2.75 tons meant it had to be mounted on a two-wheeled carriage and was used by artillery units. Even larger types of weapons were produced, such as the 9.75in mortar developed by the Americans and mounted on the chassis of the M7 self-propelled gun carriage. This particular system was actually designated as being a 'chemical mortar' with a very limited, indeed specific, role of firing either smoke projectiles or gas-filled shells. Three such monster weapons were built and deployed for use with the US 6th Army Group and Seventh US Army at Benny in France in February 1945, but were never used in combat.

All armies used mortars as support weapons to provide additional firepower and the Soviet Red Army used thousands of mortars in all calibres. At one stage in the fighting in Stalingrad during late 1942, more than 700 mortars were being operated amidst the ruins by troops such as Bezdiko, an infantryman serving with the 286th Rifle Division of the Soviet Army commanded by

Colonel Nikolay Batyuk. Bezdiko was termed an 'ace mortarman' for his ability to fire six mortar bombs into the air at the same time. At the Battle of Kursk in 1943, the Soviet Red Army deployed over 6,000 pieces of artillery and mortars to illustrate how important the Soviets regarded the mortar. Even if one allows one mortar for every ten weapons, this is still a very high concentration of mortars. Choosing Bezdiko for high praise was done purely for propaganda purposes to maintain morale among Soviet troops. There were many good mortar crews who could match and probably even out-perform his achievements.

There is an optimum calibre of the weapon which can be operated by ordinary infantrymen and this is set at 120mm and is classified as 'heavy'. Anything over this calibre can be termed 'super heavy', such as the Soviet 160mm calibre, but because such designs usually had recoil systems, were breech-loaded like a field gun and towed on wheeled carriages, these were more often operated by artillery units. Even larger calibres were developed up to 30.5cm which were as large as some field guns and were often deployed together in battery formation.

Some so-called mortar designs of the Second World War were what were termed as 'spigot' launchers. These are not true mortars in the proper sense of the term. The rear section of the spigot mortar bomb was fitted with a hollow tubular tail, which contained the propellant cartridge, which slipped over a steel rod which had been made ready by cocking the action which stored the energy of a compressed spring. On operating the trigger, the spring released the rod to fire the bomb. This method of operation places spigot mortars outside the sphere of this work and will only be referred to in passing and as necessary if the type of operation has a direct bearing on the development of a particular weapon design. The French Army, for example, used spigot-type mortars, including two versions of the *Mortier de 58T*, all in 58mm calibre. In general, though, spigot weapons followed their own path of development. The German Army also used light and heavy versions of this weapon of type known as the *Ladungswerfer*, which fired a 20cm and 38cm calibre projectile respectively. The projectile for the 38cm weapon weighed 328lbs and could be fired out to a range of almost 1,000 yards. It was used as a demolition charge to clear paths through minefields with the force of its blast and to destroy concrete emplacements, but it was never popular and withdrawn from service. At least ten of the 20cm versions of this weapon were deployed to the island of Jersey where they were to be used to fire against any landings from their positions on cliff-tops. The British Army on the other hand developed a range of spigot weapons including the Projector Infantry Anti-Tank (PIAT) along with others such as the Blacker 'Bombard' and the 'Flying Dustbin', which was a petard mortar mounted on a specially converted Churchill tank which launched a projectile weighing 40lbs out to 150 yards to destroy bunkers and other obstacles, like the German weapon.

The mortar as a defined weapon has a history stretching back many centuries to a time when extremely large calibres were common. Even in the First World War there were very large calibre mortars such as the British 6in Newton used by the Royal Artillery. These types of weapons and others inflicted casualties amid the trenches and in the open. The shells fired killed and wounded soldiers and non-combatants such as medical aid workers alike. One of these was the American-born author Ernest Hemingway. In 1918, aged barely 19, he volunteered to serve as a Red Cross ambulance driver and in June that same year was sent to the Italian Front where the war against Austria was being fought. One of his earliest duties was to assist at the scene of an explosion at a munitions factory. Only weeks later, on 8 July, he was wounded in both legs by shrapnel from an exploding mortar bomb and invalided out.

By the time of the Second World War, lighter weight mortars with more manageable calibres had been developed and were being used mainly as an infantry weapon. A number of these new, modern types had already been used in action during some of the wars which preceded the outbreak of the global war. For example, during the Spanish Civil War of 1936–1939, the Nationalist forces of General Franco, supported by Hitler and Mussolini, and the Communist-backed Republican forces each had the 5cm Ecia Valero and 8cm Mod 933 mortars available to them. The Nationalists had the heavier 12cm 'Franco' and other types of weapons were supplied to each side by their respective political supporters. During the Sino-Japanese War which began in 1937, mortars were used by troops of both sides. Conventional artillery was at a premium with the Chinese Army and the infantry divisions were equipped with eighteen to thirty mortars for fire support, including a version of the American 60mm calibre M2 weapon which was produced in China as the Type 31. This weapon was also copied by the Yugoslavian Army and used as the M5. This version weighed 41lbs assembled and ready for combat and could fire a HE bomb out to a maximum range of 2,800 yards. Switzerland remained neutral throughout the war, but even so the country still retained an armed presence along its border and its troops were armed with modern weapons including mortars such as the *Minenwerfer* M1933 of 81mm calibre. This weapon was operated by a crew of three men and weighed over 136lbs in action. It was only used by Switzerland and could fire a high explosive bomb weighing almost 7lbs out to a maximum range of almost 4,500 yards, and was comparable to all other similar weapons of this type in service with other armies at the time.

The mortar was a weapon for mobile warfare, capable of being carried ready for use by the infantry or transported mounted on specially adapted vehicles which could be called on to provide support fire. It is for this reason that mortar designs of the super heavy types such as the US 155mm T25 will not be featured in this work, which concentrates instead only on those types used by the infantry units in the many battles from Dunkirk in 1940 right the way through to the final hours as the last desperate battle was waged in Berlin in April and May 1945. Mortars were in service with the frontline fighting man in other actions such as the Normandy campaign, Monte Cassino in Italy and Operation Market Garden at Arnhem in Holland during September 1944. After the war, many of the designs of these weapons used by the Allies were to be used by their respective post-war client states. In fact many of these weapons remained in service until the 1970s, which is not a bad track record for a weapon, the design of which can be traced back to the First World War.

Chapter 1

The First Test

At 4.45am on the morning of 1 September 1939, the full might of the German Army, supported by over 2,300 aircraft from the *Luftwaffe*, launched an attack against neighbouring Poland in an operation codenamed '*Fall Weiss*' ('Case White'). Around 1.5 million German troops were committed to the attack, comprising sixty divisions, supported by 9,000 pieces of artillery and 2,750 tanks. The attack came suddenly and although it took the Polish military by surprise, it was not altogether unexpected. For several months Polish Military Intelligence had sensed a growing threat and as early as March the Polish government had ordered partial mobilisation of reserve troops. This was increased to general mobilisation on 30 August as tensions further increased. Six months earlier, as though sensing something like this would happen, Britain and France had given Poland assurances that they would guarantee the country's borders and protect its sovereignty.

Germany had been demanding access to Danzig and East Prussia through passage across the dividing strip of land known as the 'Danzig Corridor' which gave Poland a narrow access point to the Baltic Sea and separated Germany proper from East Prussia. Poland refused such access, infuriating Hitler. Polish military planners stepped up their heightened military alert, knowing that the refusal would provide Hitler with an excuse to attack. Germany had been flexing its military muscle since 1935 when Hitler repudiated the Treaty of Versailles, began a radical rearmament programme for the army and marched into the Saar region. In 1936, Hitler re-occupied the Rhineland and committed German troops to support General Franco's Nationalist forces during the Spanish Civil War. In 1938, Germany annexed Austria in the *Anschluss* and in March the following year German troops marched into Prague, the capital of Czechoslovakia. In May that year, Hitler and Benito Mussolini had signed an alliance – the so-called 'Pact of Steel'. In August, Germany and the Soviet Union signed a non-aggression pact. Britain had taken a stance of 'appeasement' and did not confront Germany directly, which led Hitler to believe he would be able to attack Poland without any recrimination. He was mistaken. Within forty-eight hours of the attack Britain and France had declared war on Germany, as had the British Dominions of Australia and New Zealand. India declared war on Germany the same day and five days later, on 8 September, so too did Canada. France also had its overseas territories in North Africa to draw troops from, such as *Goums* from Morocco. These would prove an asset later in the war.

At the time of the attack a standard German infantry division of the first wave had a manpower level of 18,000 troops, with the second wave divisions having 15,000 troops and divisions of following waves having fewer troops. A division contained all the support elements required, such as engineers, artillery, medical and supplies, with three infantry regiments being at the central core of the establishment. Each regiment had three battalions, each of which in turn was composed of three rifle companies and a machine gun company which also contained the support mortars. The machine gun company of each battalion was divided into three sections, each with nineteen men and two 8.1cm mortars and seventeen light mortars of 5cm calibre to provide fire support, along with anti-tank guns and field artillery to engage vehicles. In 1939, even 1 Cavalry Brigade had artillery support and was provided with six 8.1cm mortars. This divisional structure which was used during the Polish campaign in September 1939 would remain unaltered when, seven months later, Germany attacked Holland, France and Belgium.

As well as rearming with new weapons, the German Army had also been developing a new tactic called *Blitzkrieg*, or 'Lightning War'. In practice it was broken down into phases or elements, each of which was essentially inter-service co-operation between the *Wehrmacht* (army) and the *Luftwaffe* (air force). The first phase involved deciding on the axis or line of advance to be taken by the advancing units of infantry and armour, which had been reconnoitred by advance vehicles such as armoured cars and motorcycles scouting ahead and reporting back. Using the firepower of artillery bombardment and ground attack aircraft to provide support, the tanks moved forward with the infantry following to make contact with the enemy. Whilst the defending enemy was still reeling from the shock of the artillery and aerial assault, the armoured units attacked in heavy numbers and pushed their way through the defences. Next, having broken through the enemy positions, the armoured units moved out, circling around any points of resistance which were left for the follow-on infantry units to deal with. These armoured thrusts were designed to cut lines of communication and supply routes. In the fourth and last phase, the armour continued to advance with air support from ground attack aircraft, leaving any further remaining isolated pockets of resistance to be dealt with by the infantry. The Italians developed a similar strategy called *guerra di rapido corso*, but it was nowhere near the level of German tactics in strength of armour or co-ordinated airpower. Later in the war the Allies, especially the Soviet Red Army, would develop their own similar tactics but with far greater levels of manpower and overwhelming resources, especially armoured units.

It has been claimed that the *Blitzkrieg* tactics were not exploited to their full potential in Poland because of the rapidity of campaign. Later, against the western armies, the *Blitzkrieg* was able to be used to its maximum. German troops are often credited with having gained experience of modern fighting during the Spanish Civil War, which has since been described as being a 'live firing rehearsal' for what was to come five months after that war had ended. Whilst this was correct for some troops, the greater proportion of the German Army was not battle-tested. Germany had indeed sent 16,000 men, 600 aircraft and 200 tanks to support General Franco's Nationalist Army, but the Italian leader Benito Mussolini made a greater contribution by sending 50,000 troops between 1936 and 1939. The list of weapons sent by Italy to support the Nationalists was also impressive and included 660 aircraft, 150 tanks, 800 pieces of artillery, 10,000 machine guns and 240,000 rifles. Between July 1936 and February 1937, the Italians also sent more than 700 mortars, and by the end of the war this figure had risen to 1,500 weapons.

On 1 September 1939 the strength of the Polish Army stood at around 1.5 million troops, of which 1 million were frontline. The country had inherited a great deal of obsolete equipment and weaponry in the aftermath of the First World War and had only been recognised as a nation in its own right since 1920. Consequently the Polish Army lacked modern equipment, especially tanks. A modernisation programme to re-arm the army had started in 1937 but was far from even beginning to make any real difference. The army was organised into thirty-nine infantry divisions, including nine reserve divisions, eleven cavalry brigades, two motorised brigades and a number of other units. Artillery and support weapons comprised 4,500 field guns and mortars, 2,000 anti-tank guns and 3,000 anti-aircraft guns.

A typical infantry regiment of the Polish Army consisted of 1,900 troops divided into three battalions, each divided into platoon and company formations. Each battalion of infantry was equipped with the standard infantry weapon, the 7.92mm calibre Mauser M29 bolt-action rifle. Other weapons deployed with a regiment and in service at platoon or company level included Browning M28 light machine guns, Browning heavy machine guns, twenty-seven light mortars, six heavy mortars, two field guns and nine anti-tank guns. Much was made at the time about Polish cavalry using lances to charge down Germans tanks, and whilst the Polish Army did use mounted troops, they were supported with horse-drawn artillery and other weapons, including machine guns and mortars. The Polish Army had 210 squadrons of cavalry divided into three regiments of light cavalry, twenty-seven regiments of lancers and ten regiments of mounted rifles. These regiments were formed into eleven cavalry brigades, each with a typical strength of 7,184 officers and men, and included anti-tank and anti-aircraft guns. Additional fire support was provided by light field artillery, two 81mm calibre mortars and nine light mortars of 50mm calibre.

The heaviest mortar in service with the Polish Army was the 81mm calibre *Modziercz piechoty* wz/31 built at Pruszkow, which was developed from a design produced by the French company of Brandt. It had been the intention to equip every support company of each battalion with at least four of these weapons, but at the time of the German attack only two of these weapons were issued to each of the support companies. The *Modziercz piechoty* wz/31 mortar weighed 131.6lbs complete in action and had a barrel 4.15ft in length which could be elevated between 45 and 85 degrees to help adjust the range. It fired two types of bombs, both with a muzzle velocity of 574fps (feet per second), which was a fairly standard capability at the time. The 7.16lbs high explosive bomb had a maximum range of 3,116 yards and the heavier 14.32lbs could be fired out to 1,312 yards. Other types of ammunition could be fired

Figure 1. Re-enactor wearing Polish Army equipment of the 1939 campaign with a real w/31 mortar as used by the Polish Army in 1939.

Figure 2. Real wz/31 mortar with modern re-enactor. Note the size of the bomb (on sandbag).

by the mortar, including the standard smoke and illuminating types for screening movements and either signalling or to light up targets at night.

The light mortar was the 46mm calibre *Granatnik* wz/36 which weighed 18lbs in action and could fire a 1.6lbs bomb out to a range of 875 yards. It dated back to a design in 1927 termed as the *wz30*, but a modification in 1936 led to it being produced as the wz/36 and between 1936 and 1939 some

3,850 were produced. The barrel was 25in in length and the weapon was termed as a grenade thrower. It was issued at the rate of eighty-five per division with three to each company. It was fitted with bipod legs and could be fired at elevations between 45 and 75 degrees. A gas regulator chamber over the barrel adjusted the firing pressure to produce muzzle velocities between 98 and 360fps to adjust the range, and a rate of fire of up to fifteen rounds per minute could be achieved. The bomb was small and light, indeed no larger or heavier than the weight of a hand grenade, but the weapon could fire the bomb further than a man could throw a hand grenade, which made it a useful lightweight weapon despite its limitations.

The Polish Army fought as gallantly as any force could, but under the circumstances it was no match for the German Army, which out-numbered it and, being supported with modern weapons, tanks and aircraft, smashed anything the Poles sent against them. One of the first major engagements of the campaign was the Battle of Westerplatte, fought between 1 and 7 September, which showed that the Polish troops did not lack the tenacity or will to resist. The defences in the area were not very strong but did include four 81mm wz/31 mortars and several other types of support weapons. These were used to fire on the German attackers, who eventually had to call in Junkers 87 aircraft to neutralise the positions using air strikes. At various points across the country the Polish Army continued to fight stubbornly and bravely but by 27 September it was obvious that further resistance was futile and the Polish capital of Warsaw had no choice other than to surrender. At the time of being attacked the Polish Army comprised thirty infantry divisions, twelve cavalry divisions and ten reserve divisions. The army had almost 900 tanks, mostly small, light TKS 'tankette' designs except for ninety-five larger 7TPjw tanks equipped with 47mm guns, and whilst handled with tenacity there was little they could do against sixty German divisions with thousands of tanks and vehicles using modern tactics. The Poles suffered 66,000 killed, 200,000 wounded and over 700,000 taken prisoner. In turn they had inflicted more than 44,000 casualties on the Germans, including 10,574 killed, 30,322 wounded and 3,400 missing, and destroyed or damaged 217 tanks, but such losses could be replaced quickly and easily.

On 3 September, Prime Minister Neville Chamberlain had broadcast to the nation that, as a result of the attack on Poland, Great Britain was at war with Germany. Two days previously, Britain had given Germany forty-eight hours to respond to the demand to quit Poland, but guessing that the ultimatum would be ignored the British Army had immediately begun sending troops to France on 2 September. This was the British Expeditionary Force (BEF). By 27 September there were 152,000 troops in readiness, if and when a German attack should come. They had over 21,400 vehicles of all types and artillery support, logistics and access to reinforcements and air support. At the beginning of 1939 the strength of the British Army was 227,000 all ranks, with five regiments of Foot Guards, twenty regiments of cavalry of the line, two regiments of Household Cavalry, with mounted and armoured units, sixty-four regiments of infantry of the line along with the Royal Artillery and Royal Tank Regiment. As the situation worsened more reserves were called up, and by August the army had 428,000 men. An infantry division of the BEF had an inclusive strength of 13,600 troops organised into three brigades, each with three battalions. Support units within this structure included artillery, engineers, logistics and medical units. Mortars were available at company and platoon level, with specialist support being provided by anti-tank guns and batteries of field artillery. Two years later, in 1942, the basic structure remained unchanged in organisation but the manpower levels had increased to 18,347 troops. By 1942 and into 1943, reorganisation produced an infantry division with 17,521 officers and men in the standard triangle organisation with three battalions, and at brigade strength had 2,240 rifles, 175 light machine guns along with eighteen 3in mortars and forty-eight 2in mortars. An armoured division in 1942 had 14,964 men, with three battalions of infantry formed into a brigade equipped with eighteen 2-pdr anti-tank guns, 174 machine guns and a total of sixty-six mortars for fire support.

Following the conclusion of the Polish campaign, British and French troops faced the Germans in the west and the war entered into a period of inactivity lasting several months, during which time the troops simply sat in trenches and bunkers staring at one another along the Franco-German border. The American senator William Borah referred to it as the Phoney War, a term adopted by the British. Neville Chamberlain, the British Prime Minister, called it the Twilight War and claimed that Hitler had 'missed the bus'. The French called the period *Drole de Guerre* ('funny war') and the Germans knew it as *Sitzkreig* ('sitting war'). Whatever one chose to call it, senior French and British commanding officers knew it all rested on the next move by the Germans, and if they chose to play a waiting game the Allies would have no option but to go along. Some small localised manoeuvring on the part of the French at points on the border did take place but never any more than a token show and the troops invariably returned to their positions after completing these patrols. The Germans spent this time consolidating their forces, making good their losses and preparing for their next move. The Dutch, French and Belgian armies did nothing to prepare for an attack. The indecision would lead to catastrophe. The British Army would suffer losses but it was luckier – though only just.

Upon accepting the Polish surrender, the Germans seized large numbers of vehicles which they pressed into service and also captured vast stocks of other weapons and ammunition, much of which, such as rifle ammunition, they were able to make use of because it was the same calibre as they used in their service rifles. The light *Granatnik* wz/36 mortar does not appear to have been absorbed into service, presumably because it would have been of limited value and the Germans did not consider it worth their while to keep producing a weapon which fired such a light projectile and would almost certainly have had a small blast radius. The wz/31, on the other hand, was readily taken into service and used as the GrW31(p) for the simple reason that the Germans already had enormous stocks of 81mm ammunition suitable for use with this calibre of weapon, which was also in standard service with several European countries. This made good sense militarily and helped ease the burden of resupply to some degree during the early days of the war. In addition, the Polish factories producing ammunition for these mortars and other weapons could be used to supply the German Army and relieve the pressure on German armaments manufacturers, and thereby further increase output. By December 1939, Germany had increased its output of ammunition by more than eight times from pre-war levels. At this time Germany was producing more than 1.5 million tons of steel per month and ammunition production absorbed 400,000 tons of this output, so any additional production was a bonus.

The surrender of Poland took Britain by surprise, and the French, who had believed that Poland would be able to withstand the German Army for at least six months, were also shocked by Poland's capitulation. Several months later it would be the turn of France to shock their British allies when on 15 May 1940 the French president, Paul Reynaud, telephoned Winston Churchill to announce that the French Army had 'been defeated'. He went on to say: 'We are beaten, we have lost the battle'. Churchill responded by saying: 'Surely it can't have happened so soon?' It was a premature prediction by the French president, but one which would turn out to be perfectly correct in its assumption. Churchill's response appears to have known that such an outcome was likely, but with the strength of the French and its tank force, such an admission of defeat, less than a week after being attacked, should not have entered the minds of the country's leader. Yet here he was announcing the Germans had won.

The use of captured stocks of enemy weapons was a judicious move on the part of the German Army because it meant it was able to supplement its existing stocks. The Germans used some of these captured weapons to install in coastal fortifications along the Atlantic Wall and others were issued for service with volunteer units, internal security units and reserve troops. Another 'windfall' of weapon stocks came from the Czechoslovakian Army, which in 1938 numbered 1.5 million troops. This strength did not serve as a deterrent to Germany completing its annexation of the country in early 1939, and in the process gaining all the armaments factories which produced tanks and artillery, such as the Skoda Works at Pilsen. This factory was already producing mortars and throughout the war would remain one of the most important armaments production centres for the German Army. For example, it was producing the 81mm calibre *Minomet vz/36* which had been in production since 1936 and was intended for service with the Czechoslovakian Army. It was a standard design based on the French Brandt type, and when Germany annexed the country about 900 of these weapons were already in service. Production ceased in 1939 and those weapons in service were transferred to reserve units of the German Army, with some used during the invasion of Poland and later again in 1941 during the attack against the Soviet Union. About 150 examples were used by Slovakian troops during the little-regarded and very brief Slovak-Hungarian War between 23 March and 4 April 1939, where they were used in their primary role as support weapons.

The vz/36 fired two main types of high explosive bombs, and although it was based on the French Brandt design it could not fire the ammunition from the French Army weapons due to certain differences between the weapons. The barrel measured 45.9in in length and fired the bombs at a muzzle velocity of 720fps. The 7.2lb bomb had a range of around 3,700 yards and the heavier 15.1lbs bomb could be fired out to 1,300 yards. The weapon had the three main components, the barrel, baseplate and bipod, and weighed 137lbs in action. Each component could be carried by a member of the crew. Elevation was between 40 and 80 degrees and traverse was 10 degrees on the bipod.

Skoda also produced the 90mm calibre *Lehky Minomet* vz/17, a design which dated back to 1917. Although categorised as a 'light mortar' its weight in action of over 380lbs precluded it from being used as an independent weapon by the infantry and it had to be transported on a special trailer. The barrel was fitted to a large baseplate and tipped forward on a bracket to allow loading through the breech. Although it was an obsolete design, there were still around 212 of these weapons in service in 1938, but shortly after the outbreak of the war few, if any, remained in service because of their age and it is understood that none were ever used in action. The Skoda Works produced the heavier 140mm *Hruby Minomet* vz/18, which could fire a bomb weighing 330lbs out to a range of 2,930 yards with a muzzle velocity of 623fps, but its combat weight of over 855lbs meant it was more often deployed with artillery units.

When Adolf Hitler and the Nazi Party had come to power in 1933, the *Reichswehr* (German Army) of the Weimar Republic had 84,000 rifles, 18,000 carbines, 792 heavy machine guns and 1,134 light machine guns for an army with a nominal strength of 100,000 troops. This level had been set by the terms of the Treaty of Versailles in 1919 which forbade Germany to possess tanks, anti-tank guns and anti-aircraft guns, and similar restrictions were in place to control Germany's air power and naval force which was forbidden submarines. In 1935, Hitler announced his intentions to dismiss the Treaty of Versailles and claimed that Germany had actually destroyed many tons of weapons, including 6 million rifles, 130,000 machine guns, 340,000 tons of ammunition for these weapons and over 16 million grenades along with much peripheral materiel. This sounded impressive, but the destruction largely comprised stocks of obsolete weapons which were being replaced by more modern weapons in a rearmament programme, and in the words of Herman Goering Germany was putting 'guns before butter'. What countries such as Britain and France did not know was that secret preparations had been made which allowed German armaments manufacturers such as Krupp and Rhein-metall to produce new weapons in collaboration with companies in neutral Sweden. Officers such as General Hans von Seeckt had managed to increase the size of the army in secret by sending troops to train in other countries such as the Soviet Union. New armoured vehicles including tanks began to appear, along with specialist artillery which had been developed in secrecy, including the 8.8cm FlaK anti-aircraft gun which was used during the Spanish Civil War against tanks and would later become the dread of the Allied tank units who knew it as the '88'. No limitations had been put on standard infantry weaponry such as mortars, but nevertheless new designs of these were also introduced into service to replace obsolete weapons.

Hitler also ordered an ambitious building project to construct a defensive line along the country's western borders, from Switzerland to the border between Germany and Holland. Hitler had envisaged the idea in 1936 and the line, which was termed 'Westwall' and stretched for a distance of 390 miles incorporating 18,000 positions for machine guns, mortars and artillery, along with belts of minefields, was built between 1938 and 1940. The Allies called it the Siegfried Line and studied its development with great interest. After the fall of France in 1940, work was suspended, but construction was reactivated in June 1944 when it was realised the Allied landings at Normandy were suc-cessful. Defences were hurriedly improved, with anti-tank guns and mortars sited to engage the approaching Allies. It was much-vaunted, but in the end it did not present the Allies with any great difficulty and certainly did not stop them from crossing into Germany.

France had reacted to Germany's rearmament the following year by instituting its own rearmament programme. Several other European countries also embarked on such programmes and armament companies increased their output of weapons so that, by the time of the outbreak of war, the armaments manufacturers of the main belligerent nations were producing mortars and even supplying these designs to others who had little or no manufacturing capability. Indeed, the armed forces of all the nations that would eventually fight in the war made extensive use of mortars and over half the designs were based on the French Brandt design. Some armies would use their mortar forces to a greater degree than others, and the opposing Allied and Axis powers also supplied their respective allies with mortars. In 1939, the French Army was believed to have some 8,000 mortars of 81mm calibre alone in service with infantry units. The French had originally purchased 81mm calibre mortars from Britain in 1918, and in 1939 still had more than 2,000 of these in service. Just prior to the start of the war, the French Army had three designs of mortar in service. The first type was the Brandt mle 81.4mm 27/31 L15.6, which indicated a barrel length 15.6 times the calibre to give a barrel length of 1,269.84mm. The optimum length of a mortar barrel was set between fifteen and twenty times the calibre. The other types of weapons in service were the Brandt mle 1935 of 60.5mm calibre and the 50mm mle 1937, which was also used in static fortifications such as the Maginot Line. When the German Army launched its *Blitzkrieg* against France, the number of mortars available for fire support to a typical infantry regiment within the French Army was nine light mortars of 60mm calibre and eight of 81mm calibre. After the fall of France in June 1940, all of the weapons which had survived the fighting were taken into service with the German Army and used to bolster the armaments at various defensive locations. During the inter-war period, France had increased the strength of its army in all areas from infantry and artillery to tanks in armoured divisions; so that on the eve of declaring war in September 1939 the French Army actually had a force of 5 million reservist troops which could be mobilised and a tank force of some 3,000 vehicles of all types, while Germany had 2,700 various types of tanks available, as used during the attack on Poland.

France had also taken steps against the possibility of invasion by constructing a line of static defences along its eastern border with Germany. This was the Maginot Line, so-called after the minister of war, Andrè Maginot, who conceived the idea. The line was composed of a series of independent

Figure 3. French-built Brandt 81mm 1927/31. This was the design followed by most other mortars in the Second World War. (*SASC*)

defensive zones with overlapping arcs of fire for support, which in many cases were interconnected by subterranean lines of communications including a light railway to move troops, supplies and ammunition quickly and efficiently without being seen. Work on the Maginot Line began in earnest in 1930, with additional work being carried out in 1940. At the time of Germany's attack there were 142 *ouvrages* or works, including twenty-two major fortresses and thirty-six minor fortresses, 352 casemates, seventy-eight shelters and seventeen observation posts along with 5,000 blockhouses, stretching northward along the border with Switzerland for 450 miles. Some positions had heavy 135mm mortars in turrets with a full 360 degree traverse for firing and other locations included either 50mm or 81mm mortars in their armament.

Some turrets were retractable, which were almost undetectable when lowered, and there were twenty-one such turrets, each mounting two 81mm mortars, in the Maginot Line. Each of these could fire standard bombs out to a range of 3,000 yards at the rate of ten rounds per minute. These turrets measured 5ft in diameter and weighed 123 tons. Others were armed with a combination of hand-loaded *Mortier de 50mm mle 1935* weapons and 25mm anti-tank guns. Cupolas designated LG for *Lance Grenate* mounting 60mm automatic mortars were designed and around fifty such emplacements were built, but in the end the weapons were never installed into any of the defences. When Germany attacked France, units assaulting the Maginot Line managed to outflank the positions in some places while other points had to be attacked directly, such as that at Fermont in the La Crusnes Fortified Sector between Longuyon and Longwy, which the Germans attacked on 11 June. For more than two weeks the fighting raged, with the French defenders firing everything they had, from machine guns and 81mm mortars to artillery. Finally, the position surrendered after being subjected to an intense artillery bombardment, which included being pounded by 88mm guns and three 210mm calibre and four 305mm calibre super-heavy mortars. After France surrendered, the Germans stripped out weaponry from emplacements, including *Mortier de 50 mle 1935*, which they later used as the 5cm *Festungsgranatewerfer 210(f)*, some of which were used on the defences in the Channel Islands.

At the time of the capitulation, the French Army was using several types of mortars and a number of older designs were held in reserve stocks left over from the First World War. The lightest weapon was the 5cm *Lance Granate mle 37*, which the Germans would later use as the GrWr203(f). This had entered service in 1937 weighing only 7.27lbs in action, and was a short-range weapon firing a bomb of 1lb out to a maximum range of 550 yards. It was fitted with a small dial to allow traversing of the barrel to a small degree. Elevation was a very basic method, being done by hand using a wire looped

Figure 4. The French 50mm Lance Granate 37, used until 1940.

around the barrel which engaged in notches. The ends of the wire were attached to the bipod legs. The weapon was fitted with sights which allowed for direct fire, and a firing lever on the right-hand side by the breech was operated by the firer and this activated a firing pin to strike the base of the projectile. The mle 37 was a compact weapon and light enough to be carried as a single piece by one man, and the bombs or grenades were light enough to be carried by other members of the platoon. The Brandt 60mm mle 35 had been introduced into service in 1935 and in 1940 there were 4,900 supplied to the French Army. Its actual calibre was 60.7mm and it weighed 39.25lbs in action. A five-man crew served the weapon, which could fire a 2.8lbs bomb out to 1,860 yards or a 4.85lbs bomb to just over 1,000 yards. The Romanian Army, which would later become allied to Germany, also used this weapon and had permission to build it under licence. In May 1940, a typical French infantry regiment was issued with 200 rounds of ammunition for each weapon. The captured stocks were taken into service by the German Army and designated GrWr225(f).

The French Army had two versions of 81mm mortar in service, one of which was designated 8cm mle 44 and weighed almost 143lbs in action. It fired a bomb weighing 7.7lbs out to a maximum range of almost 3,300 yards. But it was the 8cm Brandt mle 27/31 which was the more widely-used and influential of the two weapons. The mle 27/31 weighed 131.6lbs in action and fired a 7.1lb bomb out to a maximum range of 3,116 yards or a 14.3lbs bomb to a maximum range of just over 1,300 yards. It was served by a crew of three men, and besides being the standard mortar for the French Army the weapon was also supplied to Estonia, Greece, China and the Republic of Ireland, and it was also built under licence in Romania.

Chapter Two

An Old Weapon Re-emerges

Gunpowder weapons termed as mortars had been in use with artillery as early as the fifteenth century, and a huge example was deployed at the siege of Constantinople in 1453 where it was used to batter the walls and buildings of the great city by the Turkish Army. By the sixteenth century, gun founders working in England such as Peter Baud (sometimes written as Bawd) along with Peter van Collen were casting mortars with calibres of 11in and 19in, which more than delighted King Henry VIII, who rejoiced in having a powerful artillery force. Such huge calibres meant these weapons were best used in siege operations against walled cities or castles to fire projectiles at extremely steep angles of elevation to reach over the walls at very short ranges. These types of mortars retained enormous calibres for many years and remained part of the artillery train when on campaign. Gradually the size and weight of these weapon designs was reduced to make them more mobile, which also allowed them to be more versatile in the range of targets they could be used to engage. Around 1674, the Dutch military engineer Baron Menno van Coehorn (variations in the spelling of his name include Coehoorn or Cohorn) developed a mortar which fired a projectile weighing 24lbs and was used in the siege against the Dutch city of Grave during the closing stages of the Eighty Years War. This design was much more compact than anything seen previously and light enough to be moved on a horse-drawn wagon, thereby giving the infantry its first portable mortar to use against field works.

Over the next 240 years, mortars were in continuous use by armies in various wars and some of these designs reached enormous calibres. For example, at the siege of Cadiz in Spain in 1810, the French deployed mortars with 13in calibres along with other artillery. At the siege of Antwerp in 1832, the French once again deployed gigantic mortars with calibres up to 24in. The British Army also considered adopting even larger calibres when the Irish-born civil engineer Robert Mallet proposed a built-up mortar with a calibre of 36in which he intended for use during the Crimean War. A series of events meant that the war had ended in February 1856 before his design was ready, but it continued to be developed. On being test-fired, it showed design flaws and the weapon was scrapped without firing a shot in anger. The British Army still had mortars of 13in calibre in service during the nineteenth century, and both the Confederate and Union armies used mortars of this size during the American Civil War. Some of these weighed over 7.5 tons, such as the 'Dictator' used by the Union Army at the siege of Petersburg in 1862, and were so large they had to be transported by train. Gradually the calibre of mortars was reduced again but they were still part of the artillery branch.

It was not be for another fifty years, during the early battles of the First World War in October 1914, that a real need for some form of weapon capable of firing explosive shells at short range into enemy positions was requested. By the end of that year, both sides had halted their initial sweeping movements which were aimed at trying to outflank one another and had kept the fighting mobile. The opposing armies now settled into their respective positions and started to 'dig in' and create trench systems reminiscent of the Russo-Japanese War of 1904–1905. In that war, the Russian defenders around Port Arthur had dug a series of trenches which had to be captured by attacking Japanese troops who emerged from their own trenches which surrounded the besieged location.

The trench system which developed to snake across France and Belgium eventually extended from the Swiss border to the Channel Coast, a distance of almost 500 miles, in a virtually unbroken line of defences and counter-defences. At times these positions were only a few hundred yards apart and in other places they were so close that soldiers could throw hand grenades into one another's positions. It was a stalemate and some form of weapon was needed which would allow the troops to fire projectiles further than they could throw grenades without unduly exposing themselves to enemy fire. It also had to be sufficiently compact and light enough to be moved around the trenches. Such a weapon would release the infantry from their dependency on the artillery for support, which would allow the guns to be used to fire on other targets such as enemy artillery positions, ammunition supply points and lines of communications.

General Sir John French, Commander-in-Chief of the British Expeditionary Force in France, took up the call and asked for some 'special form of artillery' which his troops could fire from their trenches to 'lob' bombs or grenades into German positions. Designs soon began to emerge, many of which were dismissed as being impractical. For example, one design hastily produced in France was based on nothing more original than a piece of 3.7in cast iron drainpipe to fire equally crude projectiles filled with explosive. Mortars which had been made in the mid-nineteenth century, including some which may have been used during the Crimean War, were rushed to the

Front, where the troops could not believe the antiquity of these weapons which they were now expected to use.

More surprising was the fact that stocks of ammunition for these weapons had been located and were of equal vintage, and units known as the 'Trench Mortar Service' were raised to use these weapons. The serviceability of these weapons was uncertain and methods to fire them from a safe distance had to be devised in the event that they should burst on being fired. At Pont de Hem near Estaires in November 1914, two artillery officers and nine gunners using these obsolete weapons were formed into a group and referred to themselves as the 'Suicide Club'. Through trial and error they managed to operate these mortars and even achieve a modicum of success, firing projectiles out to ranges of 300 yards.

Whilst these heavily dated mortars and extemporised designs were sound in principle, what the troops in the front line trenches really needed urgently was a properly produced weapon. In an attempt to produce something quickly, frustrated British troops began making extemporised weapons, which included the Second Army producing mortars using brass shell cases from a factory at Armentières. The Germans, on the other hand, were far more orga-nised and had *minenwerfers* ('mine throwers') which had been produced by the huge armaments industry of Krupp. When war broke out, the German Army had 116 medium and forty-four heavy versions of these weapons, which were categorized as trench howitzers and as such were part of the artillery. The levels of these weapons increased as the war progressed so that by the middle of 1916 there were some 1,684 of all types in service, and by the end of the war the number had increased to around 17,000 of all types.

Meanwhile in England a more promising design was being developed in the workshops at the Woolwich Arsenal in London. This was the so-called 'Twin-ing' pattern, and weapons were hurriedly sent to France in January 1915. Unfortunately they proved just as unsatisfactory as the drainpipe mortars when eight out of the eleven weapons burst on being fired in the space of ten days. Such an unreliable track record only served to produce a not un-natural reluctance among the troops to fire the weapon. Captured examples of German weapons had been sent back to England to be copied and some had been sent to France, but what was needed was a weapon design which had been properly developed and field-tested before being sent to front line troops. In 1918, the Hungarian Army was using a basic mortar design of 90mm calibre known as the *Magyar*, which was a very simple tube affair elevated and mounted on a baseplate, but worked nevertheless and around forty-eight of these weapons were issued to a division.

One person who applied himself to the task of developing a new weapon to the requirements of the army was Frederick Wilfred Scott Stokes, later to become Sir Frederick when he was knighted in 1917. He applied his engineering expertise to the problem and contrived a design which was simple and really no more than an improved idea based on the initial drainpipe design. Indeed, he personally described his idea as being, 'little more than a piece of coarse gas-piping, sitting dog-fashion on its hind quarters and propped up in front by a pair of legs corresponding to the canine front equivalent'. Stokes was born in Liverpool in 1860 and apprenticed to the Great Western Railway and took a keen interest in engineering, being involved with designing bridges for the Hull & Barnsley Railway. He later joined the Ipswich-based engineering firm of Ransomes & Rapier and became Managing Director of the company. In 1915, he was working in the Inventions Branch of the Ministry of Munitions when he devised his idea for a new mortar, which would bear his name as the Stokes mortar. Stokes later received a financial reward from the Ministry of Munitions in recognition of his work along with a royalty payment of £1 for each of his mortar bombs used during the remainder of the war.

Stokes approached the design as a means to deliver a HE bomb at short ranges fired at a steep angle to plunge into enemy trenches, where it would explode on impact. He used a smoothbore barrel, which is to say it did not have rifling grooves inside to impart a spinning action which would stabilise the bomb in flight. The base of the barrel rested on a metal baseplate and the upper end was supported on a bipod rest which could be traversed left and right. By adjusting the height of the legs the angle of fire could be altered. The projectiles were called bombs and produced as very simple cast iron cylinders filled with a HE compound. The fuse was the same type as fitted to the Mills hand grenades and fitted with a safety pin in the nose of the bomb. In the base a 12-bore shotgun-type cartridge filled with ballistite compound, a fast-burning smokeless powder, provided the propellant. By 1917, Stokes had standardised his bombs to 76mm (3in) and a bomb 12.6lbs bomb could be fired out to a range of 820 yards. The later version, known as the 3in Mk 1, fired a bomb weighing 10lbs out to a range of 2,800 yards. By the end of the war, the British Army had 1,636 Stokes mortars in service on the Western Front.

After the war, many Stokes mortars were used in the local wars of South American countries – the Paraguayan Army used them during the Chaco War of 1932, for example. The newly-created state of Poland purchased about 700 Stokes mortars between 1923 and 1926, which led to an unlicensed copy known as the *Avia wz/28* being produced. The weapon had to be abandoned

Figure 5. Diagram of Stokes 3in mortar showing its method of operation.

Figure 6. Re-enactor showing a Stokes trench mortar with 'toffee apple' bomb.

in 1931 because the bombs it fired were based on the French Brandt design and a licence to manufacture the ammunition was denied.

In operation the loader removed the safety pin from the fuse before the bomb was loaded into the barrel. The safety lever was held in place by the wall of the barrel as the bomb slid down the length of the tube. When the bomb reached the breech end it struck the firing pin, which detonated the ballistite cartridge which then propelled the bomb out of the barrel. On leaving the barrel, the safety lever was free to spring clear and the bomb was armed, allowing the nine-second time fuse to burn through. Depending on the time of flight to the target, the bomb either produced an airburst or it would explode on impact or within a second or two of landing. The loader could fire as often as required to provide support fire, or until the target was destroyed. These early cylindrical bombs were not stabilised and some were seen to tumble over in flight and land base-first. This was not a problem because the bomb would detonate whichever way it landed due to the fuse being what is termed an 'all-ways' type, meaning that it did not require the force of impact to detonate the bomb. To placate the critics, Stokes redesigned the bombs for his mortar and introduced fins to stabilise them in flight so they landed nose or fuse-first. He also made a series of other modifications to the actual mortar itself, but all the elements were there which would influence later designs.

In fact, in 1924, just as the Poles were receiving the first examples of the Stokes mortar, the French engineer Edgar William Brandt was developing his own design of mortar. Brandt began by using a Stokes weapon as a starting point. He was no new comer to designing weapons, having established an armaments factory in Paris in 1902 and produced pneumatic mortars for the French Army during the First World War. The weapon he produced was known as the M1927/31. It was 81mm calibre, which became the standardised calibre for medium mortars around the world, with the exception of Soviet Russia which adopted the calibre of 82mm. Brandt's design, was destined to succeed because it was based on Stokes' design, which was already proven in war, but with some modifications to improve it even further and take advantage of the taper-shaped bombs being developed. Once it had been perfected, Brandt's company began producing the weapon in other calibres, such as the 47mm platoon light mortar with a range of 985 yards and a company light mortar in 57mm calibre with a range of over 2,000 yards. The Model 1934 with 60mm calibre was adopted by the French Army as support weapon at company level and it was this model which was built under licence in America and became the M2 mortar in service with the US Army. Brandt also looked at the ammunition and set about developing a new and improved form of bomb. The original bombs as fired from the Stokes mortars contained the three main elements making up the bomb: the body or shell of the projectile contained the high explosive charge; the propellant in the base; and a fuse to detonate the bomb which was fitted in the head. Mortar bombs were usually cylindrical but there were some which resembled the tapering form of bombs

Figure 7. British troops during the First World War with a Stokes trench mortar about to fire a 'toffee apple' bomb.

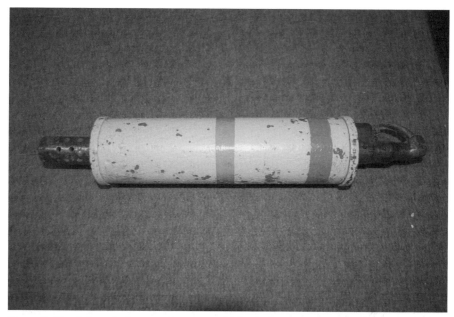

Figure 8. Bomb for a Stokes trench mortar with basic fuse. (*SASC*)

dropped by aircraft. Both types were stabilised by fins fitted to the tail and Brandt decided to improve on the latter form. His improved projectiles were demonstrated in a series of firing trials, during which they displayed a dispersion of accuracy of only 1 per cent, due to its improved aerodynamic shape and stabilising tail fin arrangement, which was completely acceptable to the military. The fins were fitted to a hollow tail pipe containing the ballistite cartridge and to this could be clipped small U-shaped containers made from celluloid and holding small amounts of ballistite compound. These were referred to as 'increment' charges and were used to extend the range of the bomb by providing more propellant force on firing, and the tapering shape of the bombs also helped in gaining extra range.

The range of mortars was expanded on by Brandt, who developed a weapon of 120mm calibre for use by infantry units at regimental level. It was also capable of being used with batteries of artillery to provide additional fire support. Brandt's other designs continued to interest more overseas armies and the M1927/31 was adopted by the French Army at battalion level, and it was also built under licence in America where it entered service with the US Army as the 81mm calibre M1 mortar. The M1927/31 comprised the three main components of the weapon design. The barrel weighed 44lbs, the bipod 40lbs and the baseplate 49lbs, to give an overall weight in action of 133lbs.

Figure 9. Detail of barrel clamp, elevating and traverse mechanism on the Brandt 1927/31.

It could fire a HE bomb out to ranges of 3,300 yards and smoke bombs to 250 yards. The main primary charge could be supplemented by up to four secondary charges to increase the range out to the maximum. A light version, known as the M1939 with a range of 2,200 yards, was intended for use at company level and to replace the 60mm weapons to provide greater fire power, but it did not proceed much further and very few are believed to have entered service.

Figure 10. Reference table for elevation and range fitted to the barrel clamp of the French Brandt 1927/31 mortar. (*SASC*)

By 1939, most armies found themselves going to war with Brandt mortars or weapons based on the French design. In the years leading up to war, the company of Brandt had expanded on its range of 81mm calibre mortars and developed a light mortar with very high power capable of firing bombs out to ranges of 5,500 yards. Another weapon of this calibre was a long-range type which could fire bombs out to 6,780 yards, but few of either type entered service before the outbreak of hostilities. The light 120mm calibre mortars were towed and could fire bombs out to ranges of almost 4,300 yards while the heavier versions could fire out more than 7,600 yards and were seen as artillery weapons. These heavy mortars attracted the attention of the Japanese and Soviet armies and were used by both sides during clashes along the Sino-Russian border, such as at Nomonham (also known as Khalkin Gol) in 1939. The Soviets were particularly impressed with the results of this large weapon and set about developing their own version, which would be used after Germany attacked in June 1941. The Germans would later capture so many of these weapons they took them into service and would even develop their own version to provide fire support.

Chapter Three

The War Moves West

When Germany finally made its next move by attacking Holland on 10 May 1940, the Dutch Army could not have been any less prepared for the weight of the assault than the Polish Army in September 1939. The war was several months old when Hitler decided to attack Western Europe, bringing to an end the Phoney War. The previous month, on April 9, Germany had attacked both Norway and Denmark. Although yielding up more weapons, the campaign also stretched the German supply lines and dispersed more troops to occupy these areas. The Danish Army was very small, with a typical infantry regiment having a strength of 3,000 men with twenty-four mortars, such as the French-built M27/31 81mm mortar which the German took into service as the GrWr275(d). Fighting during this campaign was sporadic and dispersed, and the by the early evening of the first day the campaign was over.

The conquest of Norway took longer and the Norwegian Army, with 100,000 men, put up greater resistance. An infantry regiment had a strength of 3,750 men with eight mortars for fire support, as well as field artillery which was horse-drawn. The Norwegian mortars were of the type known as the M35 *bombekaster*, of which about 150 had been produced by the Kongsberg Weapons Factory. This weighed 133.5lbs in action and the barrel was 3.8ft in length. It was operated by three men and fired a HE projectile weighing 7.9lbs, along with smoke and illuminating types. Much of the terrain was mountainous and unsuitable for motor transport. The Norwegian Army compensated for this by making great use of horses to carry heavy equipment, including support weapons such as machine guns, M35 *bombekaster* mortars and the supply of ammunition. The Germans began the invasion of Norway with an initial force of 10,000 men, including the deployment of airborne troops in a combat role for the first time. In response, Britain and France sent 15,000 troops to help the Norwegian Army, but it was obvious that the force was too small to do anything. They were not properly equipped for winter operations in the snow and the British commander, Major General Sir Adrian Carton de Wiart, reported the situation as hopeless. The Norwegian Army put up strong resistance in the country's capital of Oslo, but the Germans were able to send in more troops and supplies. By the end of May, the last troops which been with the Anglo-British force had been withdrawn and the

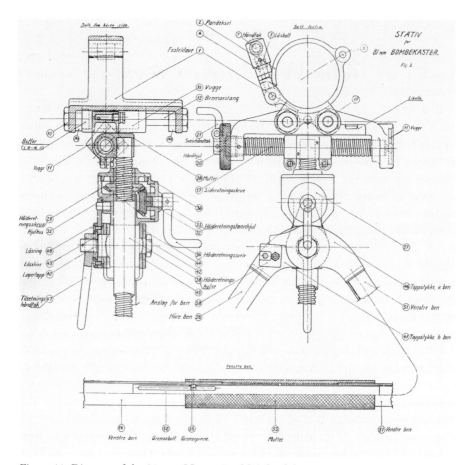

Figure 11. Diagram of the 81-mm Norwegian M37 *bombekaster*.

campaign was over, by which time Germany had attacked France, Holland and Belgium and the British Expeditionary Force was engaged in conducting a fighting retreat towards the French coastal town of Dunkirk.

Later in the war, the German Army would establish a series of *stutzpunkt-linie*, or fortified positions, in Norway from where they could conduct operations against the neighbouring Soviet Union. One of these units was the 2nd Mountain Division and one of the serving officers at the time, Major James Gebhardt, recorded the strength of armament at one such position known as Strongpoint 'Zucherhutl'. These were built along the Litsa and Titova rivers on elevated heights, from where they had commanding views. Each position was surrounded with barbed wire and minefields and had excellent fields of fire for the weaponry, which in the case of 'Zucherhutl' included, 'thirteen light machine guns (145,000 rounds), four heavy machine guns, two 80mm mortars (2,100 rounds)'. The armament from each of these could overlap with the arc of fire from other strongpoints and included light infantry guns and anti-tank guns, to provide a concentration of fire against an attacker.

The Dutch had expected an attack and although the army had been placed on a state of alert there is little doubt that the troops could not have done anything different to halt the might of the German Army's *blitzkrieg*. This was a new tactic which used aircraft cover, tanks and artillery with a combination of infantry, the weight of which easily overwhelmed the Dutch troops. Airborne landings by parachute troops seized locations and completed the element of surprise. But that is not to say that the Dutch did nothing to fight back against the attack. They did their best with what equipment they had, but it was a very one-sided battle. Holland had been neutral during the First World War and this had been respected by all belligerents, even when the fighting lapped at the country's very borders. As a consequence the Dutch Army was small and equipped to levels believed to be adequate for its own purposes. At the time of attack, it comprised a nucleus of 1,500 professional officers and 6,500 other regulars of all ranks which was responsible for training the annual intake of some 60,000 conscripts, which made up the bulk of the standing army. The length of service for peacetime conscripts was eleven months, after which time they were placed on the reserve list. Men aged between 20 and 40 were eligible for conscript service, which meant there was always a through-put of recruits. Such an organisation meant that on mobilisation the Dutch Army had 114,000 troops, and these levels would be bolstered to 270,000 with a call-up of all reserve forces. A typical infantry regiment of the Dutch Army at the time had a manpower level of 2,691 troops of all ranks. Infantrymen were familiar with a range of weapons, with each man armed with a 6.5mm calibre M-9 Mannlicher bolt-action rifle M1895 Model, with support fire being provided by weapons such as the M20 Lewis light machine gun, M08/15 Schwarzlose heavy machine gun, six 81mm mortars and a range of anti-tank guns.

The Dutch had purchased sixty mortars from the French company of Brandt in the 1920s and then obtained a permit to produce more under licence by the firm of H.H. Siderius. The version was designated the *mortier van 8*, which weighed 139lbs in action and had a barrel length of 3.65ft and could elevate between 45 and 85 degrees. It was operated by a crew of three, could fire twenty rounds per minute and was capable of firing a 7.1lbs bomb out to ranges of more than 2,950 yards. It could also fire smoke and illuminating rounds. The ammunition was produced by Dutch factories but such was the poor state of preparedness that it is estimated only 820 rounds of ammunition was available to each mortar, which were distributed at the rate of six weapons per regiment. By comparison, at the time it is believed the German Army had 6,200 such mortars in service and the French Army 8,000 81mm calibre weapons.

The Dutch Army had four military commands, each with an army corps which were centred on Amsterdam, Arnhem, Breda and Amersfoort, the corps of each one comprising a corps staff, two infantry divisions, signals and reconnaissance battalions, one independent artillery battalion and either one or two heavy artillery regiments. Other formations included anti-aircraft units, horse artillery and cyclist troops. The 4th Infantry Division was one of those units deployed and it took up positions on the left wing of the defensive position known as the Grebbe Line. This was a defensive line of trench systems which dated back to the middle of the eighteenth century and would prove no match against German weapons. The Dutch Army had some armoured vehicles, such as the four-wheeled *M39 Panserwagen* armoured car equipped with a main armament of 37mm calibre mounted in a fully traversing turret. This was a relatively modern vehicle, entering service in 1939, but it was outclassed by heavier German armoured vehicles, as were the few other armoured vehicles, when the German Army attacked. At the time of being attacked, the Dutch Army had in service between 360 and 400 81mm mortars, which was also the standard calibre in the German Army. This meant that captured examples of these weapons were perfectly serviceable and were eventually distributed to units and known variously as GrWr274(h), GrWr278(h), GrWr279(h) and GrWr286(h). Dutch armaments factories which had been producing ammunition for these weapons were seized by the German Army, and during the country's occupation these facilities were used to produce ammunition for the captured weapons and the German Army's own service weapons.

Dutch military resistance ceased after only five days, a decision which led to much criticism at the time. The decision to surrender could not have been an easy one to reach. However, with the destruction of Rotterdam the Dutch Government wished to avoid further loss of life and capitulated. At the time, many thought the Dutch Army had no fighting spirit, but this was an unfair assessment. After the fall of Holland, a number of troops managed to escape and make their way to Britain, where they were equipped and trained by the British Army. The final opinion of the Dutch military prowess is probably best summed up by Adolf Hitler himself who, on 25 May 1940, penned his personal judgement on the Dutch soldier. He wrote: 'They put up a much stronger resistance than we expected. Many of their units fought very bravely.

But they had neither appropriate training nor experience of war. For this reason they were usually overcome by German forces which were often numerically very inferior.' Whilst not exactly high praise, it is a nod of approval at their operational preparedness to fight back against a highly trained army equipped with modern weapons.

Belgium surrendered on 28 May, just as troops from the BEF and French Army were fighting a rearguard action to cover the withdrawal of their main armies to Dunkirk, from where they were being evacuated to England. Belgium, like its Dutch neighbour, had declared itself neutral in October 1936, which Germany had recognised in October 1937, but all that changed when Germany attacked on 10 May. The country maintained a standing army with a peacetime strength of 100,000 men, which increased to 550,000 on mobilisation. The army had machine guns and was equipped with anti-tank guns and some armoured cars, but being a defensive army it lacked any tank support. Divisional strength was over 9,000 men, based on the three regiment structure. For such a small size, these reduced regiments were actually well-equipped with mortars, having 108 light mortars and nine heavy mortars. The light mortar in service was the *lance grenade DBT* of 51mm calibre which weighed almost 17lbs in action and could fire a bomb weighing less than 1.5lbs out to a range of 640 yards. The Germans captured stocks of this weapon and absorbed it into service as the GrWr201(b), but its weight compared to its limited ability on the battlefield meant that it was hardly used, except probably in a training capacity to familiarise troops with handling mortars before introducing them to heavier weapons.

The Belgian troops fought to the best of their ability but after eighteen days they too were beaten, their few armoured vehicles destroyed or captured and 23,350 men killed and wounded. Sensing the Belgians were on the verge of collapse, Hitler had, three days before the Belgian Army actually surrendered, written his assessment of the Belgian soldier, stating that in his opinion: 'The Belgian soldier, too, has generally fought very bravely. His experience of war was considerably greater than the Dutch. At the beginning his tenacity was astonishing.' He continued: 'This is now decreasing visibly as the Belgian soldier realises that his basic function is to cover the British retreat.' Hitler's opinion recognised the Belgian fighting capability but his assessment of their role is incorrect because the Allies were all fighting the same retreat and the average Belgian soldier would not have been aware of any strategy at higher level.

The French Army continued to fight, but not all units fought with the same levels of commitment and this was noticed by the Germans. Indeed, Hitler was made aware of this when he wrote his assessment of the French Army which he identified as containing some 'marked differences'. He noted how in the French Army: 'Very bad units rub elbows with excellent units. In the overview, the difference in quality between the active and reserve divisions is extraordinary. Many active units have fought desperately; most of the reserve divisions, however, are far less able to endure the shock which battle inflicts on the morale of troops. For the French, as with the Dutch and Belgians, there is also the fact that they know that they are fighting in vain for objectives which are not in line with their own interests. Their morale is very affected, as they say that throughout or wherever possible the British have looked after their own units and prefer to leave the critical sectors to their allies.' In these writings, Hitler is exhibiting a degree of respect for some units of the French Army and, whilst not entirely dismissing them, he knew from personal experiences in the First World War that when an army recognises it is defeated it is impossible to motivate it. German officers commanding in the field also knew how the French could fight, as recorded by Colonel General Fedor von Bock, commanding Army Group B in the north of France facing the French Tenth and Seventh Armies. He noted in his war diary on 5 June 1940 how, 'The French are defending themselves stubbornly', and the following day he noted the situation as being 'serious' for a time.

France finally signed a formal declaration of surrender on 22 June, nine days after German troops had entered Paris, making it the last of the western European countries to surrender. Right up until the day of surrender, French troops had continued to fight. French troops in overseas theatres such as North Africa would go on to face Italian and German troops.

When Germany invaded Poland on 1 September 1939, the strength of the French Army stood at 900,000 men. On that day, orders for general mobilisation were issued to call up reservist troops which added an additional 4 million men to the ranks. In May 1940, the French Army had sixty-three infantry divisions deployed along its western border stretching from Switzerland to the coast in the north. Each division had an allocated strength of 17,000 men divided into the standard three regiments, each with three battalions, which in turn were divided into three rifle companies and a machine gun company. Each regimental headquarter had a weapons company attached to it which was equipped with six light guns of 25mm calibre and two 81mm mortars. Each regiment had nine light mortars and eight 81mm mortars for fire support and there were also at least two types of heavy mortars of 120mm calibre to provide additional fire support if needed. These were the *Mortier Brandt de 120mm Modèle 1935* and the 120mm *Cemsa Modèle 38*, which weighed almost 135lbs in action and fired a bomb weighing 36lbs out to a maximum range of almost 8,000 yards.

During the fighting, the French Army had lost 90,000 troops killed and 200,000 wounded, with the remainder of the army capitulating when the French Government signed the surrender just seventeen days after the last of the British and French troops had been evacuated from the beaches at Dunkirk. During that period, despite being outnumbered in troops and tanks, the French Army had fought on, more stubbornly in some cases than at any point in the campaign before Dunkirk. With the French surrender, the Germans inherited a modern heavy industrial base capable of producing tanks and other armoured vehicles and weapons, such as the Brandt factory which built mortars for the French Army and exported to overseas armies. They also gained the production capability of thousands of factories turning out components to complete the production of armaments and ammunition. The stocks of equipment, weapons and vehicles yielded up by the surrendered French Army was an asset which the German Army eagerly searched through to see what could be put to practical use. This included mortars with the same calibre as weapons in service with the German Army and other weaponry, with ample stocks of ammunition such as machine guns and artillery.

The campaign in France and Belgium, along with Operation Dynamo which allowed the remnants of the BEF to be evacuated from Dunkirk, had cost the British Army more than 11,000 killed, 14,000 wounded and 41,338 taken as prisoners or reported missing. In terms of equipment lost, the British Army had been forced to abandon some 63,879 vehicles including tanks and trucks, along with 20,548 motorcycles. The BEF lost around 11,000 machine guns, 90,000 rifles and of the 2,794 guns of all calibres taken to France for anti-tank and anti-aircraft roles, around 2,472 were either destroyed in the fighting or had to be abandoned because there was no room for these weapons on the evacuating ships. They were also forced to leave behind some 500,000 tons of stores, including ammunition, which was useful for the weapons left behind, including 2in and 3in mortars. The troops who returned to Britain had to be re-equipped and trained, as did those Allied troops who came out over the beaches with them, including French, Belgians and Poles. Adding to these would be Danish, Dutch and Czechoslovakian troops who were all equipped by the British Army. The French troops would later be armed using American weapons and equipment, including vehicles such as the M21 self-propelled mortar vehicle. This rearmament programme included re-equipping with mortars, and between 1940 and 1945 those British factories involved in producing armaments turned out some 40,000 units of the 2in mortar alone. In the overall scheme of things, this may not seem an unduly high output, but it was sufficient to equip British and Commonwealth forces. In addition, large numbers of 3in mortars were also produced.

With victories against Poland, France, Belgium and Holland, the German Army amassed huge stockpiles of weapons and ammunition from the surrendered armies. After assessing them, they took into service those weapons which they considered would be of use. In each case the weapon was categorised and given a 'suffix' letter to identify the country of origin. In the case of French weaponry this was '(f)' and on occasion '(e)' (*Englisch*) for British. For example, the French Brandt mle 27/31 L/15.6 of 81mm calibre became known as either the 8.1cm GrW278(f) or simply the 278/1(f), the latter term indicating that particular model had a shorter barrel length. This weapon was also used by several other European armies, including Austria, which as Germany's ally automatically made use its weaponry to Germany. Stocks of mortars captured from the Danish Army became the GrW275(d), the Czechoslovakian weapons became the GrW278(t), while some types of 81mm mortars captured from the Dutch Army became the GrW286(h) and those captured from the Yugoslavian Army became termed GrW270(j). This move of utilising captured weapons as resources was unique to the German Army. Some Allied troops used captured enemy weapons, scavenged on the battlefield to replace lost or damaged weapons, but there was never a positive move to use captured enemy weapons on the same scale as the German Army. Some types of weapons were used to supplement stocks of weapons which could be installed in coastal fortifications along the Atlantic Wall, some examples of which even found their way into strongpoints built in the Channel Islands, which had been occupied in early 1940. This meant that standard service weapons which German troops had trained on, and were familiar with their use, were left for frontline combat troops.

When Operation Dynamo was declared at an end, Britain was relieved that some 338,000 British and French troops had been saved. Winston Churchill was under no illusion that there was worse ahead and reminded people that 'wars are not won by evacuations'. The US radio journalist Ed Murrow was based in London working for the Columbia Broadcasting System, transmitting to the then neutral America, and he voiced his opinions over the airwaves, which echoed those of Churchill when he stated that to call the Dunkirk withdrawal a victory 'there will be disagreement on that point'. Churchill knew that it would take time to make good such heavy losses, and that was something which could not be wasted if the country was to withstand an invasion which seemed imminent. As the last remnants of the BEF returned to England, the army could count only 200 serviceable tanks.

On 10 June, at the height of the fighting in France, the war took a new turn when Italy declared war on Britain, with Benito Mussolini no doubt hoping to gain something from the campaign in its closing stages. On 21 June, thirty-two Italian divisions attacked but were halted by French mountain troops, and the First and Fourth Italian armies suffered heavy losses. For the Italians it was North Africa which would become their main battleground against the British Army. France had proved to be a disastrous campaign for the British Army, but amid all the chaos it did learn some valuable lessons in this first full-scale engagement against the German Army, especially when it came to the use of mortars. As a result of the fighting during the campaign, the British Army increased the numbers of mortars held by infantry battalions three-fold.

As with the Dutch, Belgian and French armies, so Hitler wrote his assessment of the British. It contained a degree of respect for the ordinary soldier but was disparaging of the officers. In his analysis of the British Army in the campaign, he opined that: 'The British soldier has retained the characteristics which he had in the First World War. Very brave and tenacious in defence, unskilful in attack, wretchedly commanded. Weapons and equipment are of the highest order, but overall the organisation is bad.' Hitler felt he understood the British soldier from personal experience, but those who had faced his army in France were a different generation – he was comparing them to the soldiers of more than twenty years previously. There were some senior NCOs who had served in the First World War and the officers, such as Lord Gort, General Montgomery and General Alan Brooke, were capable commanders; the men under them good soldiers. They fought well, especially in the rearguard actions that held off the Germans long enough to allow the evacuation to take place from Dunkirk.

One such defensive fight occurred at the harbour town of Calais, with its vital port installations. The fighting around Calais, had pushed the British inside the perimeter walls of the town, where they took up defensive positions. The Germans infiltrated some observers who took up position in the port lighthouse, in the vicinity of Courgain, and the towers of the Notre Dame church in the centre of the town. From these vantage points they were able to direct mortar fire on to British positions. An account of the action around Calais prepared in 1941 recorded how the Germans used these observation posts: 'With their guidance the fire of the German mortars was accurate and very destructive. Their mortars were so accurate that a gunner could, if he chose, lob half a dozen shells into a single window. The enemy used a great many of these weapons, bringing them up into his forward positions.' Troops were sent to clear out the observation posts, but the mortar fire continued with the same intensity, albeit less accurately. One does have to question the claim that German mortars were able to fire through windows. This is clearly an exaggeration, but to the troops on the ground that is how it must have appeared. Ron Davison served with the East Surrey Regiment at Dunkirk and remembered how the German '80mm' mortars 'had our range spot on, and never left us alone for very long'.

One of the units defending Calais was the King's Royal Rifle Corps (60th Rifles), whose positions included riflemen posted in the upper storeys of ordinary houses, offering excellent views to observe German movement. Two riflemen from the 60th were in one such building to the south of the town. They looked out onto a railway track about 300 yards distant where a passenger train stood abandoned. According to an account of the action prepared in 1941, the two men spotted 'some forty or fifty German soldiers coming up the line under cover of the train ... They were bringing up a trench-mortar. A position for the mortar had already been prepared, but to reach it the Germans had to leave the cover of the train and cross a twenty-yard gap. The riflemen were both marksmen, former competitors at Bisley. It took the Germans an hour and a half to get their mortar into action, and in those ninety minutes the two riflemen killed fourteen.' This minor action shows the determination of the Germans to get their weapon into action even at such a high cost, which indicates that the position had been identified as the spot from where it would best be able to support their advance. The two riflemen were equally determined to prevent the mortar from coming into action. The account mentions how the men left their positions just in time, as the first bomb fired by the mortar hit the building from where they had been firing. This was the nature of the fighting at the time. The situation was changing by the hour and becoming more desperate for the retreating French and British troops. In response to the German mortar fire, the best the 60th Rifles could do was to return fire with two 3in mortars, but ammunition supplies were low.

On 31 May, at Houthem on the Ypres-Comines Canal south of the city of Ypres, the defensive line around the perimeter was being held by 4th and 5th Battalions, Green Howards of 150 Brigade and 6th Battalion, Durham Light Infantry of 151 Brigade. A German reconnaissance aircraft observed their positions and almost immediately the Germans opened fire with artillery and mortars, which caused many casualties. Captain Tony Steede of the Green Howards had a stretch of 800 yards to hold with only company strength. The troops were in trenches in the open and could not be camouflaged. The Germans simply spotted them using binoculars and mortars opened fire which 'systematically plastered' their positions.

Captain Steede recalled how he was 'kneeling in one of the trenches when I got as near a direct hit from a mortar bomb as it was possible to get. My batman was on one side of me and my driver on the other; they were both killed outright and I was badly hit in the knee.' He was evacuated and taken back to England on a hospital ship. Two days earlier, at Steenstraat, Private Mons Trussler of 4th Battalion, Royal Berkshire Regiment had come under mortar fire. At 3.00pm on the afternoon of 29 May he and his section of infantry were forced to evacuate the farmhouse where they had been taking cover from German fire. Private Trussler recalled how as they left the building 'the enemy started belting us with the most accurate mortar fire I have ever seen. There were six of us, and the bombs were actually falling among us as we ran. It was impossible to take more than one or two strides before a fresh bomb sent us face down in the dirt. I saw one man with his hand sliced clean off. They mortared us all the way down the slope until we reached the breastworks at the bottom; they had been built during the 1914–18 war and formed our main defences. Our arrival was greeted by an enemy observation plane, which dropped a smoke marker almost on our heads – the signal for mortar fire.' Trussler and his comrades had obviously run down a slope which put them out of view of the Germans, but the mortar is an indirect fire weapon and, as such, can be used against fleeing troops, harrying them out of any defensive positions they may seek to take up. The British Army had gained experience in battle and faced an enemy equipped with modern weapons. Some troops, on each side, were using new weapons for the first time and saw exactly the results, which included the capability of mortars. This would be useful for later campaigns during the war.

Chapter Four

Other Developments

On 30 November 1939, whilst the eyes of the world were focussed on events in Poland and waiting to see what Hitler's next move would be, the Soviet Union brought the weight of the Red Army to bear on the neighbouring state of Finland, thereby, in effect, sparking off a war within a war. Until 1917, Finland had been an autonomous Grand Duchy within the Russian Empire, but in the wake of the Russian Revolution, which had led to the establishment of the Soviet Union, Finland proclaimed its independence. By 1919, it had emerged as a democratic state with an elected president and settled down to share an uneasy border with its communist neighbour. The situation lasted for twenty years until the Soviet Union made a claim for Finnish territories, which included the Karelian Isthmus. Finland, not un-naturally was disinclined to cede the territory and so, two months after the outbreak of war, fighting erupted between the two countries.

The Soviet Union had already flexed the muscles of its military might in the shape of the Red Army on 17 September, invading Poland to capture the eastern portion of the country, as agreed with Germany. This sudden move, coming whilst the bulk of the Polish Army was engaged against the German Army in the west, was a swift success. The attack was launched with at least 800,000 troops supported by artillery, tanks and aircraft. By 6 October, the fighting had ended with the Red Army having sustained perhaps 3,000 killed and 10,000 wounded. The Polish Army had lost 7,000 killed, 20,000 wounded and around 230,000 taken prisoner.

Seven weeks later it was the turn of Finland. The Soviet Union attacked its northern neighbour on 30 November, determined to take by force what had been demanded. This was the Winter War or Russo-Finnish War, which has rather unfairly become categorised in some quarters as a 'sideshow' and is often overlooked in the wider history of the Second World War. The fighting was overshadowed by the events which were engulfing Europe, but the war nevertheless played an important part in shaping the course of the Second World War. Germany had been a major influence on the Finnish Army for some time and even supplied it with equipment and some weapons. At the time of the war there were some fourteen factories in Finland producing weapons and ammunition, such as Ammus Oy, Wardstrom and Valtion

Tykkitehdas (State Artillery Factory), and it is believed that between them they produced almost 1 million mortar bombs of 120mm calibre and 5 million mortar bombs of 81mm calibre, of which it is estimated that 3.3 million were fired during the Winter War and later when Germany attacked the Soviet Union in 1941. The Finnish Army supplemented its winter warfare kit with items captured from the stocks of Red Army units beaten in battle, and weapons, especially machine guns and mortars, which they then used against their former owners. Among the weapons captured in the fighting were light mortars in the 50mm calibre range such as the 50-PM 38, which the Finns referred to as the 50Krh/38, the 50-PM 39, which was termed the 50Krh/39, and also the 50-PM 40, which became the 50Krh/40 in service with the Finnish Army. The stocks of captured ammunition were used with these weapons, so all specification remained the same.

At the time of the Soviet attack, Finland extended over an area of more than 338,000 square miles, with an army of around 200,000 men supported by 200 antiquated aircraft, few tanks (including thirty Vickers Light Tanks) and an assortment of obsolete artillery. The Soviet Union, by comparison, was a vast country of almost 6.6 million square miles, with huge reserves of weaponry and a virtually limitless supply of manpower to throw against Finland in a 'David and Goliath' contest. The Soviet attack was launched with 1 million men supported by 1,500 tanks (out of a total force of some 24,000 tanks) and 3,000 aircraft. It looked set to crush the Finnish Army easily.

A Soviet rifle division comprised 19,000 men, with three rifle regiments each with three battalions. Each battalion was divided into three rifle companies, with fire support being provided by a separate company equipped with machine guns and mortars. The Finnish Army had only twelve divisions, each of which comprised a combination of conscripts and reservists, and each with a typical strength of 14,200 men. Some of the weapons they had were outdated and the 120mm calibre mortars which had been ordered had not yet been delivered, but each division did have eighteen mortars of 81mm calibre. By comparison, Soviet divisions could have up to 100 mortars of all calibres in service, along with other types of support weapons.

The person responsible for introducing the mortar into the Finnish Army's inventory of weaponry is generally recognised as being Major General Vilho Nenonen. In 1924, he realised how the new 81mm Stokes-Brandt weapons would be useful in multiplying the firepower of infantry units and set about convincing others of the benefits of the weapons. In 1926, he arranged for examples to be acquired for test firings, which proved successful and led to an order being placed for seventy weapons which the Finnish Army took into service as 81Krh/26. In 1934, the Finnish armaments company of Tampella signed an agreement with the company of Brandt which granted a licence to produce mortars in return for 8 per cent of sales. Brandt also undertook to sell mortars produced by Tampella, thereby beginning a partnership between the two companies. The Finnish Army placed another order, this time for 572 81mm mortars, which were delivered between 1933 and 1939. During that time Tampella exported 579 mortars and by the time the country went to war with the Soviet Union in 1939 the Finnish Army had substantial stocks of 81mm mortars and ammunition in its armoury.

Finland was divided into nine Military Districts and the army comprised nine divisions, a cavalry brigade, field artillery and support elements. Conscripts served for eighteen months and then joined the reserve until aged 60. The Finns could never hope to stop the Red Army but, by employing guerrilla tactics, troops, known as *Sissi-Joukkeet*, were able to inflict losses out of all proportion to the size of their units and power of their weaponry. The Finns did manage to hold the border but sheer weight of numbers gave the Red Army the edge and by early 1940 the Soviets had penetrated the home defences. The Finnish Army's weaponry may have been limited but the fighting skill of the troops was second-to-none. Communications were also very basic and signals runners were widely used to convey information and messages. The ski troops wore white over-garments, which included a hood, to allow them to blend in with the snowy background.

Inevitably the Finns, with their limited resources in troops and equipment, were left with no choice but to negotiate peace. On 12 March, the fighting came to an end and the Soviet Goliath had beaten the Finnish David, but despite the disparity in firepower and equipment the fighting had cost the Red Army dearly, with the loss of 200,000 men killed and wounded, 700 aircraft lost and 1,600 tanks destroyed. Finland had turned out to be a tougher opponent than the Soviet high command had anticipated, but with the end of the war the country signed away some 16,000 square miles of territory. The Finns had lost 45,000 men wounded and 25,000 killed. For the time-being Finland remained quiet and tried to make good its losses. An inventory of weapons made in June 1940 reveals that the Finnish Army had almost 1,000 mortars of 81mm calibre in service, which had been obtained from various countries. This included Sweden, France, Italy and locally-produced types from manufacturers such as Tampella, which based some of its designs on the French Brandt weapons. For example, Tampella produced an 81mm calibre mortar which the factory called the 81Krh/32 but in service with the army was known as the 81Krh/33 and was an almost identical copy of the 81Krh/26 which had originally been supplied by Brandt between 1926 and 1927. As well as producing weapons, Tampella could undertake the repair of damaged weapons and convert captured weapons to fire other calibres of ammunition.

Finland would take up arms again and return to fight the Red Army once more, this time as Germany's ally, on 26 June 1941, four days after Germany attacked the Soviet Union. This was referred to as the Continuation War, which would last until 20 September 1944 when the Finns sought peace negotiations with the Soviet Union and then, in an about-turn, took up arms against its former German ally on 1 October 1944. During the time they fought as Germany's ally, Finnish volunteers were formed into the *Nordost* Battalion of the *Waffen SS* and the *SS Wiking* Division, both units which required their own armoured support and trucks for supply and transport. When the fighting was concluded in Finland and the campaign against France was finished, the German Army set about consolidating its territorial gains. The British Army had withdrawn back to Britain along with some French troops and prepared for invasion. New weapons were issued to replace those lost in the fighting.

The fighting in Europe may have been over, but in North Africa a new campaign was about to begin. This was desert warfare and it would prove to be a hostile and alien terrain for all troops. Largely unhindered by civilian conurbations, the armies engaged in battles that spread out to cover hundreds of miles and it would be the side which kept its army best supplied which would eventually win. The objectives were the oilfields and the Suez Canal which allowed the British to move troops more quickly. The British had maintained a presence in the Middle East since the First World War and were among the most experienced in the theatre. But that did not necessarily mean they were the best in coping with the difficulties of operating in such harsh conditions.

Benito Mussolini had seen an opportunity and ordered the deployment of Italian Army troops to Ethiopia (Abyssinia) in north-east Africa in October 1935 in an effort to create an Italian empire. In his attempt to turn the country into a colony, Mussolini sent some 400,000 troops to Ethiopia, against which the Ethiopian Army could deploy upwards of 760,000 men. However, these troops were poorly armed and trained, and against the Italians, who also used poison gas, they did not stand a chance. By May the following year, Italy had annexed the country. Italian troops had also been in Libya since 1934 and gradually built up their forces there. This now placed them either side of Egypt, where they could threaten the British Army which had bases there to protect the Suez Canal. The Italians also began to move into the coastal strip that would become Italian Somaliland, where they would eventually have some 285,000 troops, including locally-recruited units. On 4 July 1940, the Italians troops in Ethiopia moved northwards to invade British Somaliland, where they appeared to stand more chance of success than the action in France the previous month. Initially they were bolstered with more weapons, tanks and reinforcements which were sent there to replace the initial losses following the early operations. The Germans had routed the British Army from Europe and Mussolini no doubt believed he could do the same in North Africa.

At this stage in the war the Italian Army had some 2 million men under arms deployed in seventy-three divisions, of which fifty-nine divisions were infantry and the remainder designated as either armoured or specialist roles such as Alpine troops. Typically an infantry division comprised 13,500 men, with each regiment having six 81mm mortars and fifty-four Brixia light mortars of 45mm calibre which fired light HE bombs weighing around 1lb. The weapon was not popular and was described as being 'excessively compli-cated'. There were three battalions in each regiment, which meant that each one had the fire support from two 81mm mortars and eighteen light mortars of 45mm calibre, which were organised into two mortar platoons, each with nine such weapons. The structure of the infantry division which took part in the North African campaign between 1940 and 1943 was surprisingly strong in mortar support, with an independent dedicated mortar battalion. This had 435 men and was formed into three companies, each with six 81mm mortars. The Italian Army also raised a parachute division of two regiments, each with four battalions. Within this structure the support companies of the battalions were also equipped with *Modello* 35 mortars of 81mm calibre. Special Mortar Battalions, each with three companies, were created and structured within a typical infantry division, for which role each one was equipped with six *Modello* 35 mortars of 81mm calibre. The Italian Army also formed specially trained and equipped Alpine regiments which were armed with seventeen mortars of 45mm calibre and a further twelve of 81mm calibre. These units used a great number of mules to help transport the heavy equipment and ammunition in the mountains.

The terrain in Somaliland was rough and inhospitable, and mortars were the ideal weapon for use in the rugged, mountainous region, where the rocky peaks rise to heights of 15,000ft. In such conditions the fighting was desper-ately hard. Light and easily transported, mortars provided the infantry with their own artillery in these conditions and the effect of the bombs was multi-plied when they exploded on the rocky surfaces. The fighting in Somaliland would continue until late May 1941, when the Italian commander, the Duke

Figure 12. Italian Army Brixia Model 1935 as used in North Africa and the Eastern Front.

of Aosta, surrendered. On 13 September, the Italians made their next move in North Africa, attacking British forces in Egypt. The invasion came from the west, opening up a second front, and involved a force of 300,000 troops, over 1,800 pieces of artillery, 339 light tanks and more than 8,000 trucks. Air support was provided by 151 aircraft. Against this the British, under the command of General Sir Archibald Wavell, could muster 30,000 men. It appeared to be no contest. The Italians advanced 60 miles in four days before halting at Sidi Barrani on 17 September and establishing a front 50 miles wide, stretching northwards to the coast. There they remained, preparing positions and awaiting the arrival of the supplies that would allow them to continue onward. The British, meanwhile, had received reinforcements which included 148 tanks, forty-eight anti-tank guns, twenty Bofors 40mm anti-aircraft guns and forty-eight 25-pdr field guns, along with stocks of ammunition. During the campaign in North Africa from 1940 to 1943, a typical British armoured division was structured to include four battalions of infantry tasked with supporting the tanks and protecting them against anti-tank weapons. For this role the infantry were armed with mortars. Including troops in exile from occupied countries, such as France and Poland, Britain would deploy twenty-seven infantry divisions to form a field army.

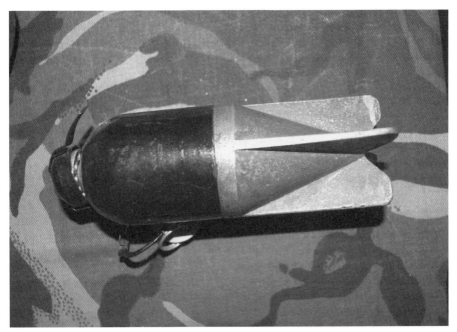

Figure 13. The small but deadly bomb fired by the Italian Army's Brixia Model 1935 as used in North Africa. (*SASC*)

Just over six weeks later, on 28 October, the Italians made their next move when Mussolini ordered an attack on Greece with a force of 160,000 men. They were badly mauled by a Greek army that had only a limited number of vehicles and was weak in anti-tank guns. The sub-zero temperatures in the Greek mountains caused great suffering amongst the Italians, who were not equipped for such conditions. They were finally pushed back by the Greeks in November. The Greek Army comprised 430,000 troops organised into five infantry divisions and fourteen mountain divisions, along with various other units. A Greek infantry division was structured in the standard way with three regiments all of three battalions. Each regiment had four 81mm mortars. Even the cavalry squadrons were equipped with four 81mm calibre mortars based on the French Brandt 27/31 designs.

In Egypt, the British sent out patrols along the Italian front to probe and look for weak points, which they finally located. On 9 December, they launched an attack and two days later had pushed the Italians back towards the Libyan border, taking 38,300 prisoners of war. By 6 January 1941, the Italians had surrendered their base at Bardia, 15 miles inside the Libyan border. The British kept up the pressure and captured Tobruk on 22 January,

Figure 14. British Army 3in mortar in action in North Africa.

and continued by following the coast road to capture Derna and Benghazi, after which Beda Fomm, Agedabia and El Agheila also fell. By early February 1941, the Italians were evacuating Cyrenaica in northern Libya and were on the verge of collapse. On 7 February, the British, with Australian reinforcements, had advanced 500 miles, taken 130,000 prisoners and captured 400 tanks, 800 pieces of artillery and huge stocks of weapons and ammunition. Realising that his Italian ally was on the point of collapse, Hitler ordered German reinforcements into Libya in the shape of the newly-created force of *Deutsches Afrika Korps* (DAK), which had been raised in February especially for the task. It was commanded by General Erwin Rommel, who had acquitted himself well during the campaign in France and was known to be an officer who got things done. The DAK was primarily an armoured formation with infantry and artillery support and it would come to gain the grudging respect of the Allies. Rommel would become a legend in his own lifetime. The infantry divisions serving with the DAK were formed into the usual three regiments, each supported by a specialist mortar company equipped with standard 81mm mortars, which were divided into three battalions and proved to be highly adaptable in many encounters.

A month after sending troops to help Italy in North Africa, Hitler was obliged to send more troops into Greece to bolster the Italians. This occurred on 6 April and was named Operation Marita. By this time the Greek Army had expanded to 540,000 troops, with more weapons and equipment. British troops had also been sent from Egypt to help reinforce Greece in March 1941. Even so, they proved no match for the Germans and after fierce fighting the campaign ended on 30 April. As usual, any weapons believed useful to the German war effort were seized, including mortars. The support companies of the Greek infantry had shown how useful mortars could be during fighting in the mountainous regions. Specialist German mountain troops would incorporate this experience into the way they handled their weaponry as they fought partisans in the mountains of Yugoslavia and Albania. To say the Italians were ill-prepared for any kind of war is not fair, because the fighting prowess and morale of the troops themselves could on occasion be extremely high. It was their lack of modern weaponry which largely led to their inability to conduct modern warfare. In June 1941, Mussolini got his opportunity to repay the support of Germany by sending troops to join in the attack against the Soviet Union. Almost 230,000 Italian troops would serve on the Eastern Front in Russia, of which more than half became casualties. By sending such numbers to Russia, Mussolini denied reinforcements and replacement weapons to the beleaguered army in North Africa, but the state of the Italian

Army was such that it could only realistically fight on one front. Trying to fight on two such widely separated fronts was beyond the capability of the Italian Army, and its logistics service would ultimately collapse.

During the campaign in North Africa, the Italians and Germans advanced eastwards on two occasions and were repulsed twice. The combined distance covered in these operations was almost 2,000 miles. This was difficult enough for the vehicles to cover but for the troops, carrying infantry weapons, it was exhausting. The British Army, along with the Commonwealth forces from India, New Zealand, South Africa and Australia, had also been forward and back twice, covering a similar distance. This backwards and forwards as each side attacked and retreated was referred to by British forces as the 'Benghazi Sweepstakes' after a term used in horseracing. The third advance made by the British forces was launched on 23 October 1942; the Battle of El Alamein, which covered over 600 miles, proved to be the decisive turning point and this time the British forces stayed forward. The battle finished on 4 November and four days later the first US troops gained their combat initiation during Operation Torch, the landings in North Africa on 8 November. This brought pressure to bear on the German and Italian rear in the west as the British kept advancing from the east. On hearing about the American landings, Rommel wrote in his war diary: 'This spells the end of the war in Africa.' It would take another six months, but he was correct in his assumption.

Over the next six months US forces fought a number of engagements such as that at Kasserine Pass and learned many lessons which would stand them in good stead later in the main European campaign. America had entered the war on 7 December 1941 following the Japanese attack on Pearl Harbor. Four days later, Hitler declared war on the USA to support his Japanese ally. The British Prime Minister, Winston Churchill, and US President Franklin Roosevelt convened the Arcadia Conference between December 1941 and January 1942, where they agreed that Germany must be defeated first. Only a year earlier, in 1940, the US Army had fewer than 250,000 men posted to some 130 bases scattered all across the country. Even when one takes into account the number of troops serving in the National Guard, the level of trained personnel was still less than 500,000 men. In September 1941, troops of the National Guard were absorbed into the army strength, by which time the overall level was 600,000 men. America had adopted a widely-held isolationist attitude towards the war in Europe and the country did not have a true sense of urgency to mobilise reservists or increase recruitment. Nevertheless, Roosevelt convinced Congress to pass the Selective Service Act in July 1941, which was a step in the right direction towards preparing the country for a war which many knew was only a question of time. Five months later, when

Japan bombed Pearl Harbor, the US armed forces were still not strong enough to conduct a prolonged war at distance, either in Europe or the Pacific. Over a period of three years, between December 1941 and December 1944, the US Army increased from around 1,657,000 men to more than 5.4 million. In fact, by December 1944 there were almost 5 million US troops serving overseas in Europe and across the Pacific and Far East.

Infantry divisions of the US Army were based on the accepted triple structure and in 1940 each of these divisions was typically formed with 15,500 troops. This was seen as being the most flexible formation to meet all 'general purpose organisation intended for open warfare in theatres permitting the use of motor transport'. This was possible in North Africa and later in Europe, but in the Far East and islands across the Pacific, with dense jungle and amphibious assaults during the island-hopping campaign, this could not be achieved. By August 1942, the divisional size had been reduced by over 1,200 men to produce a level of 14,253 troops, with some 1,440 vehicles of all types, from trucks to half-track personnel carriers. Three infantry battalions formed a regiment, with three regiments forming part of the division, which was equipped with eighty-one mortars of 60mm calibre and fifty-seven weapons of 81mm calibre, along with other self-propelled guns and towed artillery. Apart from some adjustments, this was the basic structure and size of a division which served across Europe until the end of the war.

After the Japanese attack on Pearl Harbor, followed by Hitler's declaration of war, Mussolini, not wishing to be left out, announced that Italy was also at war with the US. This was now America's fight and the country's war effort went into overdrive with recruitment increasing, vehicle production at all levels rising and newer weapons replacing obsolete types, from rifles to artillery. For example, in 1940 the American Army had only 150 M1 81mm mortars in service, such was the state of its unpreparedness, but by 1945 armaments manufacturers had produced more than 30,200 mortars of this calibre for use by the US Army, Marine Corps and for supplying to Allied forces such as the French troops fighting in Europe under General Leclerc. In the same period the Americans had 786 M2 mortars of 60mm calibre in 1940, with total wartime production rising to over 67,500 weapons. Armaments production had been moved up considerably so that by the time of the US Army's landing in North Africa (Operation Torch) a typical infantry division was equipped with eighty-one M2 60mm mortars and a further fifty-seven M1 81mm mortars for support at platoon and company levels respectively. At this time a US Army infantry division had a structured strength of 15,289 troops with three infantry brigades, each with three battalions of infantry equipped with mortars and artillery for fire support. An armoured division

Figure 15. US Army 81mm M1 (right) next to its British Army 3in counterpart. (*SASC*)

had 10,668 men in total, including three infantry battalions with support weapons including 81mm mortars. Operation Torch was the first joint Anglo-American amphibious assault of the war and involved some 107,000 men, including Free French troops and naval forces. Its success paved the way for future Allied amphibious assaults, such as Anzio, Salerno and, of course, the D-Day landings in Normandy.

Wherever American infantry was deployed to fight they would use mortars to provide fire support during attacks. This applied to all theatres of war and the weapons were either carried by infantry units, air-dropped by parachute in containers or air-landed in gliders on engagements such as Operation Market Garden in September 1944. For example, the 60mm mortar, when used with parachute troops, could be packed into a leg bag, which was tied to the para-trooper's leg by a 20ft length of rope. After exiting the aircraft the paratrooper would lower this pack so that it was on the ground only moments before he landed. With the weapon in his possession, he could immediately bring it into use as a means of providing some support to the landing zone. Unfortunately, in practice when parachuting into Normandy, many men lost their leg bags because they were torn off during the exit from the aircraft.

The American Army did field heavier calibres mortars, up to 155mm, which could fire a bomb weighing 60lbs out to ranges of more than 2,200 yards. They also had a 4.2in (107mm) M2 mortar with a rifled barrel, which was served by specially-raised Chemical Warfare Service units to provide a smokescreen. From 1943, they were provided with a high explosive round to greatly improve their utility. As such, then, this weapon along with the 105mm T13 and 155mm T25 mortars, were not standard issue to weapons companies within an infantry battalion, but were employed by separate units to provide additional fire support.

When the fighting ended in North Africa, the Allies were left free to concentrate on preparing for the next large-scale amphibious operation, the invasion of Sicily, after which it was hoped to knock Italy out of the war. Throughout the North African campaign, the Italians had continued to send reinforcements into the area, along with supplies of equipment and new weapons which could be spared. In June 1941, for example, 47,000 men arrived with tanks, anti-tank guns and ninety mortars to take the Italian strength up to 136,500 men with 780 mortars. In November 1941, the Bologna Division had 6,546 men armed with eighty-two mortars, the Brescia Division had a comparable strength, but with only twenty-one mortars, while the Pavia Division had 6,383 troops with seventy-five mortars. Between 10 June 1940 and 3 May 1943, a total of 252,839 Italian troops had been sent to North Africa and equipped with 9,314 machine guns, 1,960 tanks, 1,694 pieces of artillery, 2,322 anti-tank guns and 1,530 mortars of all calibres. The North African campaign finally ended on 9 May 1943 when the last remaining Italian and Germans forces, some 270,000 troops, surrendered. The campaign had lasted more than two years and between them the Germans and Italians lost 620,000 men killed, wounded or taken prisoner. They had also lost huge amounts of equipment, materiel and weaponry which would have served them better in the fighting on the Eastern Front. The end of the fighting in North Africa closed one theatre of operations and allowed the Allies to now turn their efforts to Europe and the conquest of Germany. Resources in weaponry, supplies, vehicles and manpower could be assigned to this aim with all urgency.

Chapter Five

A Simple Yet Effective Weapon

There are many definitions used for mortars, but perhaps the best is the one which states they are: 'Short range, high-trajectory weapons usually muzzle loaders and ordinarily small, easily transportable infantry weapons, which are used to fire projectiles in support of frontline troops, usually by means of indirect fire methods.' One of the operative words here is 'support' because it was the firepower that came through the number of rounds a mortar could fire which would provide this support. In full military parlance a mortar is defined as 'a piece of ordnance which fires projectiles at angles of elevation between 45 and 90 degrees'. Due to this arrangement, the mortar was described as being an 'outrageous novelty in gunnery'. Despite this unfair criticism, the mortar, like the larger, breech-loading howitzer, was always intended to fire projectiles 'indirectly', which is to say that the target is not always in line of sight to the crew using the weapon. Their vision may be obscured by natural obstacles such as trees and hills or man-made ones in the form of buildings. The high arc of fire allows the projectiles to clear these obstacles and engage targets. Even if the target is in plain view, the weapons are not aimed directly at it, unlike conventional artillery. Sighting onto the target is done by angle of barrel and direction of firing according to a map reference or compass bearing. Once the range has been achieved, a group of mortars will fire to suppress the target.

Although mortars are indirect fire weapons they still require aiming at a target, even when it cannot be seen. For that reason a system was developed in which the weapons were aimed by taking an alignment or bearing from a fixed point. This could be something permanent such as a large tree, prominent hill or building, or, where no such features were available, an aiming post. Once the correct angle between the line to the target and the line to the aiming post has been established and has been set into the sight, the barrel of the mortar will be pointing in the direction of the target, which may be hidden behind a hill or other feature. This setting allows the crew to open fire on the target. Once the bombs begin to land, establishing the range, then corrections can be made and the mortars will 'fire for effect'. A forward observer could relay corrections back to the firing positions and the angle of the barrel be adjusted, along with traverse left or right. The crew could also use secondary or increment propellant charges fitted to the bombs to increase the range. Aiming posts were used, indeed still are used, with medium mortars from 81mm up to the heavy calibre weapons. The posts were often carried on a vehicle along with the resupply ammunition, and even the self-propelled mortars vehicles such as the M4A1 and M21 carried aiming posts. The smaller calibre mortars did not require the use of aiming posts because the short ranges over which they were used made them unnecessary.

Mortars are termed 'crew-served weapons', which is to say they are operated by at least two men for the light designs and three to five men for the larger calibres. The lighter calibres could be fired from any convenient spot which happened to offer a good arc of fire at the time when being deployed for use. The medium mortars, on the other hand, were more likely to be operated from well-prepared dug-in positions to form batteries, not unlike artillery positions. These static positions were common in the European and North African theatres of operations, with targets being relayed to the battery sites by forward observation officers using field telephones. However, in the dense jungle conditions in the Far East and the American island-hopping campaign in the Pacific, such tactics could not always be used. The Japanese defenders holding these islands had prepared mortars positions to fire on the landings and inflicted heavy casualties on American troops as they battled to recapture islands such as Iwo Jima, Saipan and Tinian. The assaulting Americans had to battle their way forward to capture each of these positions in turn

All armies in the Second World War appreciated the value of their own mortars in battle and at the same time developed a high regard for the enemy's mortars. Because of this, the mortar was universally referred to as the 'infantry's artillery'. In fact, General Archibald Wavell believed that mortars provided 'self-support, not close support'. This was achieved by grouping mortars together in support companies, allowing them to provide immediate covering fire and to bring concentrations of high explosive shells down on to a target. Size-for-size, mortars are capable of delivering a disproportionate weight of firepower and the projectiles fired by mortars also contain a heavier explosive payload for their weight and size. The mortar was, indeed still remains, one of the most basic, yet efficient, types of weapon to enter service with any army for use in active support of infantry units. By the very nature of their design, mortars are cheap and easy to manufacture and even the ammunition which they fire is relatively less sophisticated than standard ammunition fired by field artillery. To add to the simplicity of their design, mortars lack moving parts and only require very basic field maintenance and cleaning in order to keep them in good operational condition. As with conventional tubed artillery, mortars can fire a variety of other types of ammunition in addition to high explosive shells. These include smoke to help conceal infantry movement across open spaces, and illuminating shells to light up potential targets at night.

Mortars are also quite simple weapons to operate with a common basic form, which allows them to be used easily and quickly. Even in the middle of a battle they can be quickly repositioned to fire in any direction. For that reason soldiers tend to take them for granted. Some designs used in the Second World War were produced which were more complicated than necessary and termed in some quarters as being 'over-engineered', such as the German 5cm and the Italian Brixia 45mm calibre *Modello* 35. As the war progressed there were attempts to simplify these designs and some types were taken out of service, such as the German 5cm, which was discontinued at company level in 1942. There was not much which could be done by way of simplifying mortars – they were already very simple. The barrel, for example, not having any rifling, was a plain drawn tube which required little in the way of machining apart from producing a better finish to the standardised calibre. Other older, obsolete designs were taken out of frontline service, usually to relieve the pressures of production, but some of these still had a use and were retained for training purposes

A number light mortars, such as the French 50mm *Modele* 37, British 2in and Japanese Type 98 were one-piece weapons, but the medium mortars of 81mm calibre usually comprised a smoothbore tube, enclosed at one end to serve as the barrel, a baseplate, on which the lower end of the barrel rests, and either a bipod or tripod to support the upper or muzzle end of the barrel. The barrels of most infantry mortars are smooth bore, which is to say they do not have the rifling grooves found on conventional artillery that serves to stabilise the projectile in flight after firing by imparting a spinning action. Mortars compensate for the lack of rifling by the fitting of fins to the tail end of the bomb, which acts like the fletching on an arrow and imparts a degree of spin. The smooth bore allows the mortar to be loaded quickly and simply by dropping the bomb down the barrel from the muzzle, as with a flintlock musket. The medium mortars of 81mm calibre are fitted with fixed firing pins in the breech so that when a bomb slides down its own weight makes it heavy enough to strike the firing pin and fire it. This means all the loader has to do is insert bombs into the barrel and there is no need for a firing mechanism. Smaller calibre weapons, such as the British 2in mortar and Japanese 50mm calibre Type 89, fired bombs that were quite light and did not have sufficient weight to operate in the same way as the larger calibre weapons. In order

Figure 16. The French 50mm Lance Granate 37. The compact design meant the weapon could be simply and swiftly deployed. (*SASC*)

to fire these lighter weapons, a firing mechanism in the form of a simple lever was usually fitted. On being fired, the bomb creates a relatively low pressure in the breech chamber, which is sufficient to discharge the bomb but is significantly less than a field gun. This means the tube making up the barrel could be made thinner than the barrel of a piece of artillery. The bomb projectile could be produced with a thinner casing, which allowed a greater explosive payload to be carried. Cast iron is inexpensive, but when supplies of even the most basic metals became difficult for Germany and Japan to obtain, the factories producing mortars bombs had to resort to using even lower grade metals which allowed the higher grade metals such as tungsten (wolfram) to be used for anti-tank shells or tools. Even then, this did not cause any significant problems because of the low pressures exerted on the bombs during firing. Like all weapons, mortars create recoil forces when fired, but unlike other weapons, such as field guns, the action is directed downwards through the baseplate and into the ground, which eliminates the need for complex recoil recuperating and buffer mechanisms.

By the time hostilities started in earnest, all the major armies of the world had mortars of either 50mm or 81mm calibre in service at platoon and company level respectively, with some countries having a combination of both in

Figure 17. The small grenade bomb fired by the Lance Granate 37. (*SASC*)

order to give the maximum range of flexibility to infantry units on the battle-field. Some countries developed models with very large calibres, with the intention of providing greater flexibility in their mortar forces, but these were larger than the standard infantry units in the field could handle. This meant that in 1939 there were three categories of mortar types. First were the light models with calibres between 50mm and 60mm for use at platoon level, with at least one weapon of this type being held in this unit of command, though often this was exceeded with sometimes as many as three being held on strength. These were usually supported by the firer's hand when used with the small integral baseplate resting on the ground. Exceptions in this range included the Polish Army, which used a 46mm calibre weapon in the shape of the *Granatnik wz/36*, and the Italian Army, which had the 45mm calibre Brixia Model 35 for use at platoon level. The Soviet Red Army used two types of mortars which were of exceptionally light calibre. The first was the 37mm 'spade mortar', the second was the 40mm light mortar which weighed 18lbs in action and fired a bomb weighing around 14oz out to a range of over 750 yards. The spade mortar was in service between 1939 and 1941 and weighed around 3.3lbs. It was a very simple weapon comprising a barrel attached to a broad baseplate which resembled a spade (hence its nickname). It was carried by one man who aimed it by pointing the barrel in the direction of the target and firing. The 1.5lbs HE bomb was finned and had a range of over 250 yards. The firer also carried fifteen bombs ready for use in a special belt. The spade mortar was used during the Russo-Finnish War of 1939–1940, but it did not perform particularly well according to some accounts. In effect it was like a grenade-thrower similar to the Polish wz/36, but much lighter and simpler to operate. It was still in service at the time of the German attack in June 1941 and a number were captured and put into service under the designation 3.7cm *Spatengranatwerfer* 161(r). This weapon was obsolete after 1941 and taken out of service in 1942.

The second category of infantry mortars were the medium calibre weapons, which were usually standardised around 81mm calibre, and were found to be best for serving at support company level. These were built up from three components; the barrel, bipod barrel rest and baseplate. The baseplate could be of a variety of shapes, either round, square or rectangular. The lower or breech end of the barrel rested on this component and the recoil force of the weapon was directed through this and into the ground. The Japanese went slightly larger with their medium mortar design and developed the Type 94 and Type 97, both of which had a calibre of 90mm and were used at support company level.

Finally there were the smooth bore heavy weapons of 120mm calibre, or greater, which often incorporated an integral baseplate, and for which reason were usually fitted with simple wheeled carriages for mobility. This feature also permitted these weapons to be manhandled by the crew in an emergency situation. The US Army, for example, had mortar designs up to 155mm calibre and the British army had a 4.2in (107mm) calibre design, but these were used by Chemical Warfare Service (CWS) Smoke Companies and the Royal Artillery respectively. The British Army introduced the 4.2in mortar into service proper from 1942 onwards but some may have been used on trials as early as 1940. They were intended to augment artillery and some were deployed to specialist units of anti-tank guns to provide fire support. In use with infantry units it was allotted to Brigade Support Companies provided by machine gun battalions from 1942 until the end of the war. It was usually towed on a wheeled mounting and served by a crew of five men. The barrel weighed 90lbs, the bipod 64lbs and the baseplate 177lbs to give a weight in action of 331lbs. It could fire smoke bombs weighing 22.5lbs to screen movement and HE bombs weighing 20lbs out to a range of 4,100 yards.

One unit which was equipped with the 4.2in mortar was the 43rd Division (Wessex) that operated in a specialist reconnaissance role. There were a number of battalions and regiments making up the division including 8th Battalion, Middlesex Regiment, nicknamed 'The Diehards', who fired the mortars and received target indications from Forward Observation Officers

(FOOs). The No. 1 of the five-man crew commanded the weapon. The No. 2 laid the mortar, set the range and direction on the sight unit, then aligned it with an aiming post set in front of the weapon. The other members of the crew loaded bombs into the barrel for firing and prepared the bombs for firing. A good crew worked to a method and knew the routine, and the sights were constantly being checked and adjusted throughout any fire mission in order to maintain range and direction. An extract from the War Diary of 8th Middlesex Regiment from July 1944 during the Normandy campaign records how these weapons were used to provide fire support. At the time the battalion was located near the town of Tourville on the Caen-Villers Bocage road and had sustained many casualties from German mortar fire. The diary records: 'Our 4.2-in mortars really came into their own as Arty [artillery] ammo is in short supply due to bad weather and storms in June. We have plenty of mortar bombs to break up incessant counter-attacks.' The diary relates that 1,200 bombs were available, and with each one weighing 20lbs this represented a total weight of 24,000lbs or 10.7 tons of ammunition readily available for firing. Indeed, so intensely did they fire their weapons to provide support that the men reeled around because their balance had been upset by their eardrums being pounded by the constant noise.

Other armies also used the 4.2in (107mm) calibre mortar, such as the US Army which had in service the M1 which weighed 330lbs in action and could fire a HE bomb weighing 24.5lbs out to ranges of almost 4,500 yards. During the initial phases of the invasion of Sicily, US troops were attacked by German tanks, and 4.2in calibre mortars, known by US troops as 'four-deucers', were directed against them, even though HE bombs of mortars did little or no damage to tanks under ordinary circumstances. A blast from one of the bombs landing close to a tank might damage the tracks to render it immobile but it would not destroy it. The 4.2in mortars used against the German tanks on Sicily managed to use the force of the blasts from their bombs to cause the tanks to divert and push them into a position where they were engaged by anti-tank guns. The US Army also developed the 105mm calibre T13, which weighed 190lbs in action and could fire a 35lbs HE bomb out to ranges of 4,400 yards. The Germans developed three mortars designs with the slightly smaller calibre of 105mm, which were termed as 10cm NbW35, NbW40 and NbW51 and categorised as *nebelwerfer* or smoke throwers. The Germans used these to fire high explosive bombs as an interim between the medium and heavy mortars. In the case of the NbW35, it weighed 231lbs in action and could fire a HE bomb weighing 16.25lbs out to ranges of 3,300 yards. The NbW40 and NbW51 weighed 1,763lbs and 1,435lbs respectively. The NbW40 fired a bomb weighing 19lbs to a range of 7,000 yards, whilst the

NbW51 fired a slightly heavier bomb of almost 20lbs out to 6,500 yards. They also developed the heavier 120mm calibre Gr.W42, which weighed almost 630lbs in action and fired a HE bomb of almost 34.5lbs out to a range of over 6,600 yards.

Each of the main belligerent nations had their allies which they supplied with weapons and uniforms. America had the assistance of the Filipino forces until the Japanese captured the Philippine Islands. The British Army could count on troops from Commonwealth forces across the British Empire who rallied to support Britain in its fight against Germany and Italy, and later Japan. Canadian troops, for example, began arriving in Britain by December 1939 and these had to be equipped with additional equipment in standard service. Australia, New Zealand, South Africa and India added to the numbers, and although these countries had limited facilities to produce equipment they did their best. Whilst the German Army was busy stripping weapons out of the occupied countries such as Poland, France, Norway, Holland and Belgium, to be put to use for training and being incorporated into defensive positions, Britain was negotiating with America for supplies. In March 1941, Britain and America concluded an agreement known as Lend Lease whereby weapons, vehicles and equipment would be supplied. Under this programme, America prepared to supply more than 5,200 M1 mortars of 81mm calibre. When the Soviet Union was attacked in June 1941, almost half this figure was diverted to supply the Red Army. Fortunately the Americans also supplied the ammunition for these weapons, because the Soviet mortars being of 82mm calibre meant that the ammunition for these was too large to be fired from the American-supplied weapons. Before the first batch of America's Lend Lease contribution arrived, the Soviet Army had been using 2in and 3in mortars supplied by Britain. Eventually Soviet factories would make good the losses but as an interim these weapons were better than nothing. The British Army had already been equipping troops which had escaped from European countries just prior to their fall. These armed forces included Dutch, Czechoslovaks and Poles and became known as 'armies in exile'. For the most part these were armed and equipped by Britain, but the burden began to lessen with much-needed aid from America, firstly in the form of Lend Lease, which then increased after the US entered the war in December 1941.

Troops of the Indian Army fought in North Africa against the Italians and Germans, and later against the Japanese. At the start of the fighting in the Far East, India had 900,000 men under arms, but by the end of the war this level had increased to 2,065,000, and they suffered over 102,000 casualties. The strength of an Indian division was around 10,000 men, who were armed with standard service weapons of the British Army including 2in and 3in mortars.

To start with there were difficulties concerning sufficient supplies of ammunition for the mortars, which led to serious problems, especially during the Japanese invasion of the British territory of Hong Kong in December 1941. For example, 2nd Battalion, 14th Punjab Regiment had received training in the use of the 3in mortar and crews had fired a limited series of rounds to familiarise them with the weapon, but lack of ammunition meant that such training was kept to a minimum. Supplies of 3in mortar ammunition did begin to arrive in November 1941, but only sufficient stock to allow seventy rounds for each battalion despite the growing threat of a Japanese attack, which eventually came on 7 December. In some instances Indian troops only fired their mortars just hours before engaging in battle. The situation with 2in mortars was not much better at the time but reorganisation did lead to improvements much later. Canadian troops fought in the early stages of the war in the Far East and suffered from the same problem of ammunition shortages for their mortars during the Hong Kong campaign. Indeed, at the time of the Japanese attack there was a crisis with supplies and only 300 rounds of ammunition was held in stock for 3in mortars in the whole of Canada, and none could be spared for units deployed to Hong Kong.

The main British force in the Far East was the Fourteenth Army, which came to number a million men between 1944 and the end of the war. Australian troops were engaged in the fight against Japan and the country's army expanded to 727,700 men, of which 396,661 served overseas; they lost 61,575 men killed, wounded or taken prisoner. The Indian mountain artillery regiments were usually equipped with at least twelve howitzers of 3.7in calibre and in some cases up to sixteen of these weapons were deployed along with a number of 3in mortars. A division might have up to thirty-six 6-pdr anti-tank guns, but the Japanese deployed few tanks and those that were sent into action were light and poorly armed, which led to a reconfiguration of weaponry within the anti-tank regiments. The number of guns was reduced and the numbers of 3in mortars increased to thirty-six weapons. These were in addition to those used at battalion level, with six mortars in a platoon. This level of firepower gave the Anglo-Indian troops the advantage, especially in actions fought at close range. The Fourteenth Army could be supplied by air drops and because mortar ammunition weighed less than artillery shells and took up less space in an aeroplane, greater quantities of this type of ammunition could be delivered compared to artillery shells. The Japanese troops came to respect and fear the ability of the British mortar crews in providing fire support to their troops during engagements. For example, in the night attack against Mogaung a mortar barrage of 1,000 rounds was fired to support the assault. Using mortars in such a way was unusual but it was necessary to provide fire support because they had no field artillery. The expenditure of ammunition, with each HE bomb weighing 10lbs, represented an overall weight of 10,000lbs (4.46 tons), which was a considerable load for men and mules to carry over long distances through jungle and over mountainous terrain. The Japanese returned fire using their mortars, but the British were able to advance behind their own barrage and the attack was successful.

During the campaign, the British Army used thousands of mules as pack animals to transport heavy loads on the extended operations deep into the jungle and mountainous areas of the region. Sergeant Joe Adamse served with 16 Brigade and was in charge of a mortar section in the support platoon of 22 Column. He remembered how the mules in his unit were used to carry the heavy Vickers machine guns and the two 3in mortars and ammunition. If the Japanese respected the British 3in mortar, the British troops in return knew how formidable Japanese mortars could be when used in the jungle and how the bombs would inflict severe casualties. Phillip Sharpe, also serving with 16 Brigade, was a signaller with 45 Column in the Reconnaissance Regiment, and he recalled how on the forty-fifth day of a patrol deep into Burma his unit was halted in order for a village to be scouted for signs of Japanese troops. Sharpe recollected how machine-gun fire opened up and suddenly: 'Mortar bombs began to fall behind us; one exploded as it hit the treetops overhead with a deafening roar. Branches, leaves and shrapnel showered down around us. A hail of bullets whistled fan-wise overhead, ricocheting off the trees. One mortar bomb exploded between our two signal mules and one was torn apart by the blast, the blood splattering all around us.' The barrage destroyed much equipment, including the radio, and the Japanese approached the British positions. Fortunately they were able to withdraw to avoid further losses.

Chapter Six

Ammunition

The earliest types of ammunition fired from mortars were developed to become complete rounds with the propellant charge in the base, the main explosive filling or 'burster' charge in the body and the fuse cap in the nose. This all-in-one design made it much easier for the infantryman to load and fire without any need to prepare the ammunition beyond arming the fuse. By the time of the Second World War, mortar ammunition had been developed to produce three types which were considered sufficient to meet most situations. The smoke bomb, usually with a white phosphorous filling, was used to screen the movement of infantry and vehicles moving over open ground when they would be exposed to view by the enemy and the smoke would prevent accurate fire being brought to bear on them. Illuminating ammunition was for use at night to light up the battlefield to reveal any movement being made by the enemy under cover of darkness. The third type was the high explosive bombs which could be used against the enemy in the open, even if sheltering in trenches, and these bombs were also effective against soft-skinned targets such as unarmoured vehicles, including trucks carrying stores.

Technically speaking the type of ammunition fired from mortars is more properly referred to as a 'bomb'. There are several theories surrounding the reason why they are called bombs, with the widely accepted version being that the ammunition, fitted with fins and aerodynamically streamlined, look like bombs. The fins serve to give the bombs stability during its trajectory flight to the target, and the aerodynamic shape ensures that the projectile will always land nose-first, to impact on the ground and detonate. The fins of a mortar bomb were attached to the tail tube, which also contained the propellant charge to fire the projectile, and were secured in place by either rivets, small screws or even a 'stabbing' which pierced the metal of the surfaces and secured them together in a crimping action. Unfortunately, the stabilising fins of a mortar bomb were prone to damage in combat and if not handled correctly during transportation or loading. To prevent damage to the fins, mortar bombs were usually transported in tubes, which were made from steel, pressed cardboard or even woven wicker as supplies of raw materials such as steel became in short supply. Once removed from this packaging the fins could become bent or broken, and whilst this would not affect the function of the bomb it could affect its accuracy and stability in flight. The British and Americans favoured six fins, as did the Swiss Army at the time, which kept production very simple. The American Army experimented with a design of bomb fitted with fins which were spring-loaded and deployed on being fired. This was the M45 bomb, which was stored with the fins folded, but it was discovered that over time the springs weakened and the fins did not deploy properly on being fired. In March 1940, the replacement M56 type bomb was introduced with the conventional fixed-type of fin arrangement. The German Army and the Soviet Red Army used bombs which had multiple fins – up to ten or more – which may have improved stability in flight but also meant they were more delicate and easier to damage on the battlefield.

The mortar bomb derives its ballistic propulsion by means of the propellant cartridge filled with a compound known as ballistite, which is located in the boom section of the tail fin assembly. As it strikes against the firing pin the primer is fired, which in turn ignites the main propellant charge. To increase the range of the bomb, additional propellant charges, sometimes called increment charges, can be fitted to the tail section of the bomb prior to firing. These are semi-circular discs of celluloid which are simply and easily clipped on to the bomb by springs or rubber bands and can be removed to reduce the range of the bomb. This means of adjusting the range is usually reserved for mortars of 81mm calibre and over, as well as adjusting the angle of the barrel by means of the elevating wheel on the bipod mount. In the case of the lighter calibres, adjusting the range is achieved by means of simply altering the angle of the barrel during firing. Although sounding rather

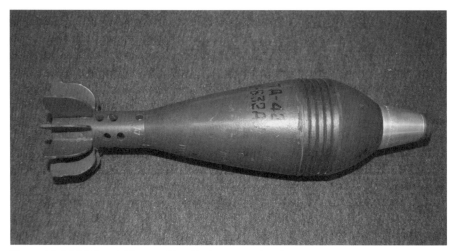

Figure 18. Multi-finned 82mm calibre Soviet mortar bomb. (*SASC*)

ambiguous, it was a tactic that worked rather well during the war and experienced mortar crews could be quite accurate.

The usual form of operating all infantry mortars is that, when loaded, the bomb is inserted into the barrel tail-first, an action which allows it to slide down the length of the tube under its own weight. With medium and heavier mortars, the downward motion of the bomb allows it to impinge on a fixed firing pin to detonate the primer cap, which in turn sets off the charge train to the increment charges. Of course this type of firing action only works well if the weight of the bomb is sufficient to permit it to strike the firing pin with enough force. Thus it serves best with 81mm mortars up to 120mm calibres which have long smooth barrels. In the case of lighter calibre mortars, such as the Italian 45mm Brixia, and the more standard 50mm to 60mm calibres, including the British 2in, which had shorter barrels and lighter bombs, it was necessary to employ a simple form of 'trip' or 'trigger-type' mechanism to initiate firing. In such cases the firing mechanism served as the action on a spring-operated firing pin to impinge on the primer charge in the base of the bomb to fire the projectile. This was because the lighter bomb lacked the weight to hit a fixed firing pin with any force sufficient to cause detonation. In such cases the firing mechanisms were usually operated by the firer with a lanyard or a simple tripping lever. In 1944, the British Army trialled a version of the 2in mortar known as the 'Weston' which used an automatic re-cocking firing mechanism. Fitted with a curved baseplate, it resembled the Japanese Type 98 'knee mortar', but it proved unsatisfactory and the idea was abandoned.

Mortar bombs are slightly smaller in diameter than the calibre of the firing tube in order that they may slide down its length more easily when loading. The unfortunate drawback to this is that it does not create a gas-tight seal or obturation and the slight space between projectile and the inner wall of the barrel produces a phenomenon during firing known as 'windage' – a loss of propellant gases escaping forward around the bomb. This leads to a drop in pressure, which in turn affects the range. During the war there was not much which could be done to correct this effect and it was something which had to be accepted and adjustments made so the bombs would fall on target. To the soldier firing the mortar in battle, such matters concerning ballistic science were of little interest. He simply wanted to know that the weapon would fire and cared little about its internal functions, unless it failed and he would complete the immediate action drills to correct the fault, which might be a broken firing pin.

Compared to the shells fired by conventional tubed artillery, mortar bombs can actually deliver a greater payload of explosive to a target in relation to the

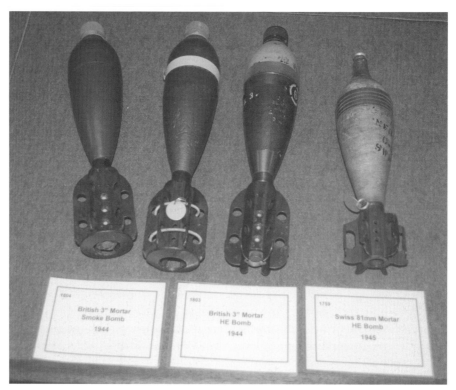

Figure 19. Range of British 3in mortar bombs, with Swiss 81mm calibre bomb (right) for comparison. (*SASC*)

relatively small calibres. On arriving at the target area, the bomb drops in a steep angle to give a good all-round blast radius when it deflagrates, throwing out metal fragments of its brittle cast iron casing in a full 360-degree sweep. When a bomb lands on a hard surface this 'killing radius' is increased, but when the bombs land on a soft surface, such as sand, a great deal of the blast can be absorbed. When directed against so-called 'soft targets', such as ammunition dumps and vehicle parks, the mortar could still give a good account of itself. Being readily available to infantry, commanders on the spot could call on their own support mortars at, say, company level, whereas artillery support had to be called for by channels higher up the chain of command. Among the Allies, it was the Soviet Red Army which came to use mortars with the most devastating effect. American and British forces made extensive use of mortars, but their respective artillery organisations were flexible enough to be used either independently or in conjunction with heavy mortars in the range of 120mm and the larger calibres.

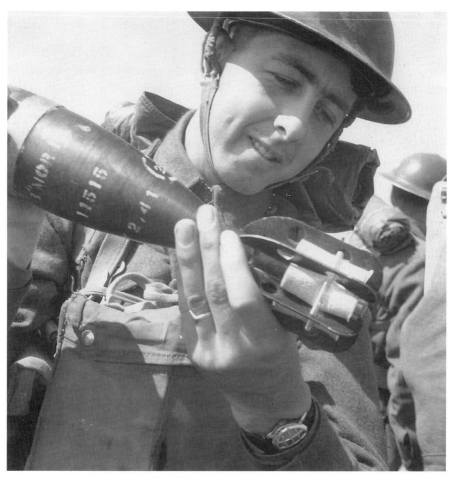

Figure 20. British 3in mortar bomb fitted with secondary increment propellant charges to increase range.

Mortars did not always have to be directed onto target areas by forward observers as in batteries of artillery with field guns, but accuracy was improved if the fall of shot was reported so that adjustments could be made to engage targets better, especially if they were moving (infantry advancing over open ground, for example). It was the simplicity of the mortar which made it an ideal weapon to arm guerrillas, partisans or resistance fighters in their war against occupying German troops.

Mortars have always been used to fire ammunition of three types to meet all situations and, indeed, that still remains the case today. The smoke bomb was usually based on the action of white phosphorous which ignites spontaneously on contact with air and produces huge amounts of smoke. Such bombs were used mainly at short ranges and were intended to screen infantry movement across open ground. The enemy would have been alerted to the fact that movement was taking place because of the smoke screen, but being unable to see exactly what was happening they weren't able to engage directly, only speculatively. White phosphorous bombs also had a secondary effect because they could ignite inflammable material. For example, on one occasion when German infantry ran into wheat fields during the Normandy campaign, Allied mortar platoons fired smoke bombs and the white phosphorous set fire to the dry stalks of wheat. The Germans were forced to run out and were shot down as they emerged. On another occasion a mortar platoon opened fire on a German convoy moving along a road using a ratio of one white phosphorous bomb to every three HE bombs. The phosphorous set fire to the trucks and as the crews attempted to run away they were caught in the open by the HE bombs. Illuminating projectiles are used at night to light up the battlefield to reveal any movement being made by the enemy, who may be attempting to attack under cover of darkness. In this type of ammunition the carrier body also contained a small parachute to suspend the bomb so that the light was cast in all directions. Finally, there was the high explosive bomb which was used against infantry in the open, trenches, soft-skinned vehicles such as trucks and other vulnerable targets such as petrol or ammunition dumps.

The smoke bomb was the most basic type of bomb, requiring no specialist fuse, and could be used to mark targets to be engaged by heavier field artillery at longer ranges or to mask the movement of troops. Various smoke-producing compounds had been tried in the First World War, such as sulphur

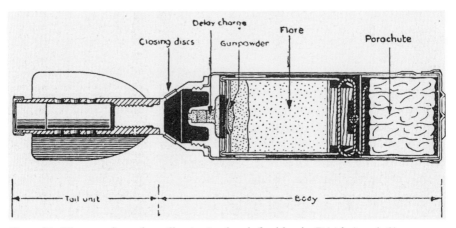

Figure 21. Diagram of parachute illuminating bomb fired by the British Army's 2in mortar.

trioxide, chlorosulphonic acid, oleum and titanium tetrachloride. But none of these matched white phosphorous (WP) for producing voluminous amounts of smoke. By the time of the Second World War, most armies had settled on WP as the most effective compound for producing smoke quickly. A small explosive burster charge was placed under the fuse, which detonated when the bomb landed. The nature of WP meant that it ignited when it was exposed to air and produced dense clouds of white smoke. Sometimes dyes were added to the compound to produce coloured smoke which could be used to identify targets to be attacked by air strikes or artillery bombardment. The smoke bombs were usually short-range projectiles and if there was a wind they were fired up-wind so that the breeze would blow smoke across the frontage of the target to screen troops advancing to attack. If the wind was blowing towards the attackers, the smoke bombs would be fired behind the enemy positions so that smoke drifted forward to obscure the advancing troops. The light calibre mortars through to the heavy calibre types could fire concentrations of smoke bombs to conceal river crossing and tank movement.

The illuminating bomb incorporated a small parachute to suspend it in mid-air to light up the surrounding area and expose any movement being made by the enemy using the cover of darkness. Illuminating projectiles had been used as early as 1866, with designs such as the 'Boxer Parachute Light Ball' fired from an 8in calibre mortar. It was relatively simple to produce illuminating bombs for mortars in most calibres, except for the small types such as the Italian Brixia *Modello* 35 of 45mm calibre. The compound most

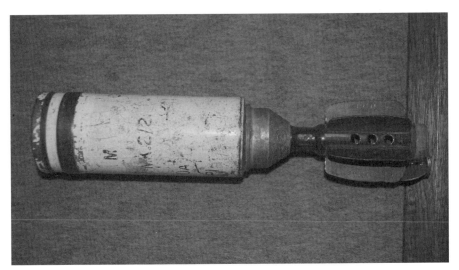

Figure 22. Smoke bomb projectile for the British Army 2in mortar. (*SASC*)

widely used to produce the illumination was based on magnesium. The luminosity and duration of the burn time depended on the calibre of the bomb. Smaller bombs such as 2in or 50mm tended to burn quickly with low illumination, while 81mm calibre bombs produced much better effect and the 120mm calibre bombs were best of all, with intense illumination. The bomb is made up with the magnesium compound contained in the forward section and a parachute in the rear section. After firing, the nose cap is ejected and the magnesium compound is ignited. The parachute is ejected by a spring and the body of the bomb is attached to it, so that when it deploys the parachute suspends the illuminating bomb like a floating beacon. Troops caught in the open were instructed to stand still because their movement would attract attention. The tactic worked in some cases but experienced troops could distinguish objects even among the harsh shadows cast by the synthetic light. The intensity of the light projected is expressed in terms of candela or candle power. In the larger calibre mortars the light emitted could be up to a million candles. The duration of burn time also depended of the size of the illuminating bomb, from a few seconds to almost a minute.

The last type of mortar bomb was the high explosive projectile, which could prove lethal at all ranges. Post-war records show that 80 per cent of US Army casualties are attributed to mortar fire, 10 per cent to artillery and the

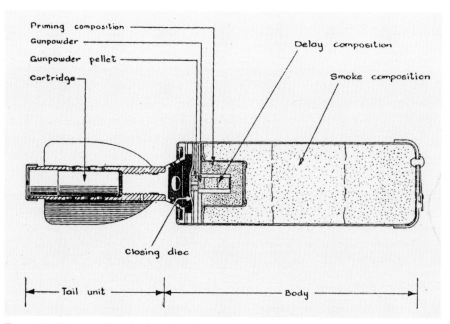

Figure 23. Diagram of smoke bomb used by the British Army 2in mortar.

remainder to 'other causes' including small arms fire, mines and blast. Even if one allows for some overlap between 105mm artillery shells and 120mm mortar bombs, this figure still represents a very high percentage of casualties being attributed to a single category of weapon. British reports from Normandy support this analysis and credit German mortars with being responsible for inflicting at least 70 per cent of all casualty figures. Indeed, the historians John Keegan and the late Richard Holmes, in their book *Soldiers; A History of Men in Battle*, refer to the mortar as 'one of war's major casualty-producers'. which they continue to explain by saying how it 'gave the infantry-man the ability to reach out to the other side of the hill'. In other words, as an indirect fire weapon, there was no protection for soldiers taking cover behind a wall or hill against the lobbing effect of mortar bombs as they dropped down. An assessment of casualty figures for the North Africa campaign reaches a similar assessment and concludes that 75 per cent of casualties were caused by a combination of mortar and artillery fire. On the Eastern Front it is generally accepted that 61 per cent of all casualties in the Red Army between 1944 and 1945 were caused by mortars. This is probably due to the fact that mortars were being used in street fighting as the Soviets advanced through cities where these weapons could be used along with other infantry weapons.

The most obvious reason why there should be such high casualty figures is due to the heavy amount of explosive filling carried as a payload in a mortar bomb in ratio to its size. Typically, an artillery shell weighing, say, 55lbs would have an explosive filling of between 8 and 10 per cent of its overall weight. This means between 4.4lbs and 5.5lbs of explosive filling made up the weight of the shell. A mortar bomb, on the other hand, carried up to 20 per cent explosive filling, which meant that an 81mm calibre bomb weighing 14.3lbs as fired by the Japanese Type 99 would contain over 2.8lbs of explosive. The larger the calibre, the heavier the bomb, and thus the more explosive filling to create a blast and throw metal splinters out to greater ranges. For this reason some units engaged in operations would place priority on the use of mortars. For example, during the early stages of Operation Market Garden in September 1944, British airborne forces fighting around the bridge at Arnhem emphasised the importance of airlifting mortars and troops. Until such times that heavier weapons arrived, the parachutists fighting as infantry used their mortars to provide fire support and engage German mortar positions. In response, the Germans used their mortars because they were lighter to move around the battle areas which developed as groups broke up into pockets. From the very start of the fighting, the Germans had more artillery but gradually they placed more dependence on their mortars. Six out of the seven Jeeps used by B Company of the 1st Battalion, Border Regiment were destroyed by mortar fire. Even Major General R.E. Urquhart commanding the 1st Airborne Division had his Jeep destroyed by mortar fire. It was not just against the British that the Germans brought these support weapons to bear. In the area of Mook near Nijmegen, one German unit with 3,400 troops had 130 machine guns and twenty-four mortars to engage units of the American 82nd Airborne Division which had landed there by parachute.

The steep angle of firing mortar bombs, in turn produces a steep angle of descent, up to 70 degrees to the horizontal plane. Because of this effect, as the projectiles impact in the target area they will detonate to give a more even, all-round blast effect, throwing out shell splinters from the bomb casing as it deflagrates. Some of the blast will obviously go upwards but much will be concentrated low, which is effective against troops in the open, plunging into trenches and engaging light unarmoured trucks. Apart from throwing shell splinters in all directions, when a HE mortar bomb is detonated it explodes with a report that creates a phenomenon known as overpressure, which can burst eardrums and cause concussion to disorientate troops in the blast range. Major General Roy Urquhart witnessed this phenomenon during the battles around Arnhem. Another strange phenomenon he witnessed was when Major Peter Waddy was struck down by the blast of mortar bombs. He recalled how when the officer was later picked up to be treated for possible wounds there was not a mark on him – he had been killed by the blast effect alone. To minimise the effect of the blast, troops would often open their mouths in an attempt to equalise the pressure and try to press their bodies into the ground in order to reduce the possibility of wounding. If the troops were in trenches they were better protected from the lethal effects of an exploding bomb which could throw metal splinters from the casing out in a radius of up to 100 yards. The plunging effect of the bombs meant they could hit trenches directly from above. If the bombs fell through a tree canopy they could strike branches to detonate and send a cascade of metal splinters and shards of wood to shower down.

Craftsman Joe Roberts, serving with the Royal Electrical and Mechanical Engineers, remembered the effect of a shell blast even when in a trench: 'You had to be careful when you put your head out of a trench because you built them to protect you from shell burst. Once I was standing in my trench observing the scene when a shell landed near by – I didn't hear it coming: the explosion crushed my chest, it took my breath away and I spent days spitting up blood.' The effect was the same whether it was an artillery shell or a mortar bomb, and Joe Roberts was fortunate to survive being caught in the blast. Captain Bill Towill, of the 3rd Battalion, 9th Ghurkha Regiment, was attached to 111 Brigade when his unit came under mortar fire from Japanese positions at night during an intense tropical storm. During an inspection of the defensive positions he remembered how: 'In all the noise and tumult it was quite impossible to hear a mortar bomb falling, as we usually managed to do. As I stopped and raised my head to look over the top, two mortar bombs landed absolutely simultaneously and within 3 ft of me, on either side of the trench. I felt a crashing, stunning blow across the top and back of my head, as if some giant hand had taken a mighty swipe at me, which knocked me violently face down in the mud at the bottom of the trench.' Like Joe Roberts, he too was lucky to escape with only concussion.

Not all troops reacted so stoically when coming under fire by mortars. The first instinct for an infantryman when under fire, whether from a machine gun or artillery barrage, was to get down and seek cover. In Europe, during the Normandy campaign, German mortar crews exploited this reaction and developed a method of firing which bracketed the area where troops were lying down to take shelter. Such fire provoked another response – making the troops stand up and often run back the way they had come. This in turn exposed them to machine-gun fire and further mortar fire. US troops had certainly been taught very different tactics to use when being fired on by machine guns and mortars during their thirteen weeks on the Infantry Basic Training Course in America. These drills were later reaffirmed on the live-firing training grounds in Britain such as Brauton Burrows in Devon. US infantrymen were taught to keep advancing, because it is more difficult to engage a moving target than a stationary one. Indeed, General Omar Bradley's Headquarters emphasised how important is was to 'keep moving if you want to live'. Troops were told to resist the temptation to fall flat to avoid enemy fire. Such instructions make sense to those issuing them – they are not in the firing line. The men doing the fighting tend to do anything to prevent them being killed, and if that means lying flat, then so be it. The reaction was not unique to the Allies or to Europe; troops of every army in all theatres from Russia to North Africa and Burma did exactly the same thing.

It was the rising numbers of casualties caused by mortar fire which led to a call by the Allied armies requesting a reliable form of equipment to be developed which would provide countermeasures to mortars. Forward Observation Officers (FOOs) could detect the sound of firing or flash of artillery when it was fired, and this allowed them to work out the range and location of artillery positions. This same technique was applied to mortars. One unit in particular became very adept at this form of calculating the range and location of German mortars. During Operation Martlet in late June and early July 1944, as the 49th Division was approaching Tessel Wood in the Odon Valley,

Lieutenant Colonel Hart Dyke, Commanding Officer of the Hallamshire Battalion serving with 146 Brigade of the 49th Division, recalled: 'The 49th Division produced a wonderful counter-mortar and counter-battery organization which swiftly dealt with any enemy fire. Compass bearings, called shell-reps and mortar-reps [mortar-reports] were sent in by each company as soon as an enemy gun or mortar fired on us, or by the battalion on our left. Our artillery officers, Harold Sykes or Mickie Carter, whichever was in the line with me, then did the necessary wherever we could get a trisection. By the time we left the area we had definitely obtained fire superiority over the enemy.' What is being mentioned here is a very simple means of locating a target from three points of detection known as 'triangulation'. Each location makes its 'mortar report' to give direction and approximate range. When three such reports have been made from separate locations, lines are drawn on the map and the target should be at the point where the lines cross. Counter-fire can then be directed against that spot, with adjustments for range. It was a method which worked and as troops became more proficient at reporting, the accuracy improved. But if the firing could be recorded and reported sooner, then response time would be improved and this would come with electronic equipment.

Scientific technicians applied themselves to producing a radar set specially designed to detect a mortar bomb in flight. The British Army led the way in this field, and by using centimetric anti-aircraft radar sets they were able to produce mortar detecting equipment towards the end of the war. Electronic signals were transmitted which calculated the parabolic trajectory of the projectile, and using these signals the operator could estimate the point from which it had been fired with an accuracy of 35 to 40 yards in around twenty seconds. This meant that by the time the third or fourth mortar bomb had been fired, the countermeasure radar had identified the location and references could be passed to artillery or heavy mortar batteries to engage the enemy position. During Operation Shingle, the Anglo-American landings at Anzio on 22 January 1944, the Allies had had been virtually unopposed. Despite this, they remained within their established beachhead and did not begin to break out until late May. During this time the German artillery and mortars fired into the landing area to cause some 73 per cent of the total 46,000 casualties of the entire operation. Anti-aircraft radar sets were used in an attempt to track the trajectory of shells to try and pinpoint the locations of the German weapons in order to engage them with counter-battery fire. At one point the radar sets were detecting so many shells they were able to locate German mortar positions within 50 yards, which was sufficient for the artillery to respond. A similar method was also used in the Pacific war and in

October 1944 a unit of Corps artillery was locating 9 per cent of its counter-battery targets using radar. In November 1944, British 21st Army Group experimented with GL III gun laying anti-aircraft radar sets and discovered that operators could use the equipment to locate a mortar position to within 50 yards.

A more sophisticated counter-mortar battery organisation was later created by the Allies and by the time of Operation Veritable, the battle for the Reichswald in February 1945, there were some thirteen such units operational. The organisation was used to its fullest during this operation, with properly linked observers, detecting radar and counter-fire units working together for the first time on the battlefield. Operating the equipment required great skill and operators had first to locate a mortar bomb in flight and calculate where it would land, and by plotting the trajectory backwards they could work out the general location of the mortar. It required quick response times and calculations had to be done in an instant and relayed to the weapons. Counter-battery fire using either artillery or mortars could be directed against the location to neutralise it. The signature noise of a mortar being fired is not very loud compared to conventional artillery, and the speed at which the sound was detected also depended on the calibre of the weapon. The speed of sound is not a constant and depends on prevailing conditions such as temperature, air pressure and moisture. In ideal conditions sound will travel at 1,125 fps in dry air at a temperature of 68°F, it also depends on the size of the weapon being fired. For example, 3in calibre mortar would register 13,123 fps while a heavier 4.2in calibre mortars registered sound waves travelling much faster at 19,685 fps. This technology would be used after the war to develop more sophisticated and sensitive equipment, leading to modern mobile battlefield radar systems.

Chapter Seven

Barbarossa and the War on the Eastern Front

The war was twenty-one months old and still going in Germany's favour when Hitler ordered Directive 21, or 'Case Barbarossa', the code to launch the attack against the Soviet Union. At 3.15am on 22 June 1941, a single gun fired to signal the start of the attack, and from that moment on all other aspects of the war seemed to be of secondary importance. Japan may not have known of Hitler's intentions, even though they were allied in the Pact of Steel. Italy, on the other hand, knew full well and Mussolini had ordered the deployment of Italian troops to support the attack. Hitler had never attended a military academy but he had a general grasp of strategy. His generals presented him with information and made suggestions, but the final decision was his. If things went according to plan he took the credit, and if they went wrong he laid the blame squarely on the shoulders of others. He believed that the Soviet Union was so corrupt that all 'we have to do is kick in the front door and the whole rotten structure will come crashing down'. In ordering the invasion of the Soviet Union, Hitler plainly had no idea of the scale of problems which lay before the German Army in this vast country. The distances covered with apparent ease during the early phase of the campaign led the Nazi leader to believe that another victory lay ahead. German intelligence assessments said the Red Army had an armoured force of some 24,000 tanks but believed that most were obsolete and would pose no problem to the modern tank forces and anti-tank guns of the German Army. The Red Air Force would be overwhelmed as their outdated aircraft were shot down by the *Luftwaffe*'s modern fighters such as the *Messerschmitt* Bf 109. Within days of the start of the campaign, German troops had advanced deep into the Soviet Union with seemingly nothing to prevent it.

The scale of the attack was unprecedented and included 3 million troops in 146 divisions supported by three air fleets with over 1,800 aircraft. Seven armies and four Panzer groups with 3,580 armoured fighting vehicles, 7,184 pieces of artillery, 600,000 other vehicles for transport and liaison roles and 750,000 horses were committed to the attack. The tank force included 1,440 PzKw III and between 517 and 550 PzKw IV, with the remainder being made up of 410 older PzKw I and 746 PzK II tanks, along with a number of PzKw35(t) and 3(t) tanks. This was *blitzkrieg* on a grand scale and it looked as though the tactics which had worked so well in Western Europe and against Poland would soon add another victory to Germany's list of conquests. Some believed that Hitler may have taken on an enemy that was too strong for his armed forces; after all, the Soviet or Red Army was estimated to number around 3 million. Other strategists thought the Soviets were fatally weakened by the army purges of 1937–1938 in which Stalin had ordered the liquidation of around 35,000 officers, robbing the Red Army of 90 per cent of its generals.

The strength of the Soviet Red Army in June 1941 stood at 5.5 million troops in all branches, and it was equipped with 91,400 pieces of artillery and mortars, but only 2,780,000 troops and 43,872 pieces of artillery and mortars were deployed in the west to oppose the German attack. The German Army also had the combined support of a further 1 million troops from their allies of Hungary, Italy and Bulgaria, who between them had almost 12,700 pieces of artillery and mortars. A typical Soviet tank division at the time had eighteen mortars to provide fire support and forty guns for anti-tank and anti-aircraft defence. A German panzer division at the time had thirty mortars and seventy-two guns for anti-tank and anti-aircraft roles. A Soviet artillery division had more than 100 HM38 heavy mortars of 120mm calibre organised into a special mortar brigade to provide fire support, something to which the Germans did not have any equivalent. The fighting was fierce from the very beginning and just three weeks after the start of the German attack the Soviet Army had lost 8,000 tanks, along with 9,427 pieces of artillery and mortars. Soviet losses continued to mount over the following weeks. For example, in the fighting around the Smolensk Pocket between July and September 1941, the Red Army lost 486,000 men killed, wounded or taken prisoner along with more than 1,300 tanks destroyed and 9,920 artillery pieces and mortars destroyed or captured. Six months later, having lost much ground by withdrawing before the Germans, the Soviet Army had lost hundreds of thousands more men killed and captured, while vast stocks of weapons and ammunition were either destroyed or captured in the fighting. Tank and aircraft losses were huge and available artillery and mortars were reduced to fewer than 22,000. The number of 120mm calibre mortars captured by the Germans and the stocks of ammunition was so great that they were pressed into action against their former owners without any need to convert them for service. Indeed, the Germans were so impressed with this weapon they even developed their own version, known as the 12cm calibre GrW42.

Soviet armaments production fell, and resupplying all the lost weapons and re-equipping new divisions would take time. The Soviet Red Army realised

Figure 24. British-built version of the Soviet HM38 120mm heavy mortar.

that it needed time to regroup, rearm and reorganize if it were to stop the advance of the German Army. Once that had been achieved it would then be in a position to push the invader back. One strategy used was the age-old tactic of giving up ground, called 'scorched earth'. This meant that the Red Army left nothing behind in its wake that would be of any use to the Germans. This placed a heavier burden on the already over-stretched supply lines of the German Army, which had to bring everything forward as it pressed ever-deeper into the country. In turn this tactic bought the Soviets time to gather sufficient forces and weapons to mount a counter-offensive. The number of troops killed, wounded and taken prisoner continued to rise, and the levels of weapons, tanks, vehicles and aircraft lost was a testimony to the ferocity of the fighting. After six months of seemingly unstoppable advance, the German Army finally ground to a halt in the temperatures of -30 degrees in the outer suburbs of Moscow. The Red Army seized the opportunity to mount a counter-attack on 5 December, and with 720,000 men supported by 670 tanks, 5,900 pieces of artillery and mortars and over 400 rocket launchers they pushed the Germans back almost 150 miles to recapture the city of Smolensk. Moscow was safe for the time being but far from completely secure.

The weight of the German attack made the Soviets recognise that their industrial centres were at risk of being captured or destroyed, and a massive effort was put into moving hundreds of factories, especially those producing weapons for the army, thousands of miles to the east beyond the Ural Mountains. This placed them beyond the range of German bombers and armaments production could begin to replace losses. Britain and America sent military aid in the shape of tanks, aircraft and trucks. Britain even produced versions of the HM38 120mm calibre mortars to replace the losses incurred by the Soviet Red Army. This was a special production because the British Army had never at any time used such a weapon. Once the relocated Soviet factories were firmly established, weapon production began to increase and the losses of the early months were made good. The factories also produced heavier calibre mortars such as the M43 160mm, which weighed 1.15 tons in action and could fire a HE bomb weighing 90lbs out to ranges of over 5,600 yards. These heavier mortars were breech-loading weapons and regiments equipped with them were formed to serve with the Artillery Armies. Armed with this combination of weapons, such artillery units would unleash massive bombardments to smash the German forces with weight of firepower.

The Soviet Army was not entirely without combat experience, having sent advisers to support the Republican forces during the Spanish Civil War. They also sent weapons including tanks, aircraft and artillery which were operated by small numbers of Soviet troops. The Red Army had also been engaged in

Figure 25. Soviet armaments factory producing mortars.

frequent border clashes with Japan, such as the incident near the Soviet port of Vladivostok in 1938. On another occasion, between March and September 1939, Soviet and Japanese troops fought a series of engagements in the area of Khalkin Gol which saw Soviet troops defeating the Japanese. At the time, Japanese troops were engaged in fighting in China and these border clashes with the Soviet Union had caused thousands of casualties to both sides. The short but bloody Russo-Finnish War between November 1939 and March 1940 had also added to the combat experience of the Red Army. In all of these engagements, troops and weapons had been tried and tested in combat, which would later be used during the opening engagements against the German Army in 1941. After the early German successes, which led to the capture of vast stocks of Soviet equipment, some of these weapons would end up being used against them.

To support the German attack against the Soviet Union, the Hungarians and Romanians deployed 44,000 and 358,000 troops respectively. Hungary had been Germany's ally in the First World War as part of the Austro-Hungarian Empire. Romania had been Germany's enemy in the First World War, but in the war against the Soviet Union, Romanian troops fought alongside German troops as allies and together with Hungary they deployed between them a combined force of 3,445 artillery pieces and mortars. Now, as Germany's allies once more, they deployed between them a combined force of 3,455 pieces of artillery and mortars. Only a year earlier, in June 1940, Romania had been militarily undecided and as the main oil-producing country in the region it was seen as strategically vital to both Germany and the Soviet Union. Germany had pre-empted any move which may have been made by the Soviet Union by infiltrating troops into the country on the pretext of training the Romanian Army. In reality they were deployed to safeguard supplies of oil to the German Army, along with those of Hungary and Bulgaria. By September there were some 18,000 so-called German instructors in Romania, whose presence was still being explained as necessary to help modernise the army. By November all pretence was dropped when the country's Prime Minister, Ion Antonescu, signed the Axis Pact which allied Romania with Germany.

The Romanian Army incurred heavy losses throughout the Soviet campaign, especially during the fighting in the region of the Ukraine and Crimea. In June 1941, Romanian troops were serving as part of the German Eleventh Army to capture Sevastopol, where a force of over 720 mortars was engaged in the action. By August 1942, the Romanian Third and Fourth armies were engaged in the fighting at Stalingrad. Germany supplied the Romanian Army with a large proportion of weaponry, including anti-tank guns and mortars.

Some of these weapons were from captured stocks of French mortars such as the Brandt 60mm Model 1935. They also received 81mm Brandt Model 1927/31 and Model 1939 weapons. Romania had an armaments industry capable of producing machine guns, small arms and mortars, including a 120mm calibre weapon called the Model 1942, built by Resita. This weapon was towed on a two-wheeled carriage and had a barrel length of 6.1ft and weighed over 617lbs in action. It could fire HE bombs weighing 35.25lbs out to ranges of 6,780 yards. The Romanians also had pre-war permission to build the French 50mm Brandt Model 1937 mortar under licence, which the Romanian Army later used during fighting against the Red Army. This weapon weighed 7.27lbs in action and could fire HE bombs weighing 1lb pound out to ranges of 550 yards. It also used other mortars, including a version of the German heavy calibre 12cm GrW42 produced by Rosita.

The Hungarian Army made its first military incursions of the war when it independently invaded neighbouring Slovakia in October 1939. The country was pro-German but it did not commit itself to joining the Axis Pact until November 1940. The Hungarian Army had a peacetime strength of just 80,000 troops and like Bulgaria and Romania its weaponry was outdated and it was largely reliant on horses to move artillery. With German support the country very quickly expanded its strength and, in June 1941, put troops in the line alongside German and Romanian forces during Operation Barbarossa. Within a month of the campaign starting, a unit called the Hungarian Rapid Force (also known as the Hungarian Rapid Corps), comprising about 40,000 men, and composed of troops from VIII Corps and 1 Mountain Brigade, advanced deep into the Donets Basin alongside the German Seventeenth Army, where it took part in the Battle of Uman which garnered thousands of Red Army prisoners. The Second Hungarian Army was composed of nine light infantry divisions, each of which had two infantry regiments with their structure supported by obsolete tanks such as Panzer MkI which only had machine guns. The Second Army was deployed to Stalingrad, where it was given the task of holding a front line some ninety miles long. When the Red Army offensive of January 1943 was launched it very quickly penetrated the positions held by the Hungarian troops, who fell back leaving some 148,000 killed, wounded and captured, losing much weaponry in the process, including more than 400 mortars of 81mm calibre.

As Germany's ally, the Hungarian Army continued to fight and at the beginning of February 1945 still had 214,000 men deployed. Some elements had surrendered in late 1944 but other units staunchly held out by the side of Germany until the end of the war in May 1945. The fighting would eventually cost 300,000 casualties and many taken prisoner. The weapons used by the infantry units included mortars such as the 50mm calibre 39M produced by FÈG and a Hungarian-built version of the GrW36, known as the 36M. Hungarian factories such as Dimàveg, EMAG and Bàmert produced 81mm calibre mortars such as the 36/39M and 100mm calibre 41M. The 36/39M was a Brandt design weighing 187lbs in action and capable of firing a HE bomb weighing 9lbs out to ranges between 55 and 4,700 yards. Also produced by the factory of Diòsgyor, these weapons were in service at the rate of four per battalion. Factories also produced the 120mm calibre 43M, a Hungarian version of the German GrW42, and Dimàveg also produced the 90mm calibre 17M, a version of the Czechoslovakian *Lehky Minomet* vz/17 produced by Skoda. Hungarian troops also used captured Red Army weapons such as the 42M 82mm calibre mortar.

Finland also joined the war against the Soviet Union, referring to their renewed fighting as the Continuation War; to some this was seen as an ideal opportunity to complete unfinished business from two years earlier. The Finnish Army deployed more than 300,000 troops with over 2,000 pieces of artillery and mortars. Although Finnish troops did fight in other areas of the Soviet Union as volunteers serving with the *Nordost Battalion* of the *SS Wiking* Division, the main effort was concentrated in the northern area of the Karelian Isthmus. Here they took up positions and maintained the blockade whilst the Germans occupied positions to the south to besiege the city of Leningrad in an operation which would last almost 890 days or twenty-nine months between September 1941 and January 1944. Volunteers from occupied countries such as France and Belgium came forward to fight in SS units known as the *Charlemagne* and *Wallonien* divisions respectively, and even some Dutch nationals volunteered to serve in the *Nederland* division, all of which had to be equipped with standard service weaponry. Germany's European allies, Romania, Hungary and Finland, also had to be supplied with weapons, which placed a strain on support services to keep them equipped and production output was also under pressure.

Germany's other European ally was the state of Bulgaria, which had also been allied to Germany in the First World War. The Bulgarian Army consigned itself to conducting anti-partisan campaigns in Greece and Yugoslavia and by mid-1944 comprised twenty-one infantry divisions and two cavalry divisions, along with other units such as two frontier brigades. It was an outdated force, despite its armoured brigade being equipped with over 120 tanks of French and German design, and lacking in modern anti-tank guns. It made extensive use of horses for transport of supplies, which actually turned out to be ideal for use in the mountainous regions in sweeps against partisans. Bulgaria joined the Tripartite Pact on 1 March 1941 but did not participate in the fighting in the Soviet Union. The army was equipped mainly with German weapons, including mortars of 8cm and 5cm calibre, along with a range of captured enemy weapons which were considered sufficient for the role in which the troops were engaged.

The Germans had been halted and forced back when some units were within 15 miles of Moscow. Weather conditions had played a significant part, with sub-zero temperatures preventing ammunition and fuel from being brought forward. The Red Army pressed forward and by February 1942, under the command of General Georgi Zhukov, had pushed the Germans back between 90 and 180 miles in some places. The Soviets kept up the pressure, hoping to surround the Germans as they consolidated at Kharkov. Sensing the threat, the Germans moved first and attacked. By 23 May they had surrounded their attackers and captured 200,000 men and killed a further 70,000. Hoping to seize the initiative again, Hitler ordered the 6th Army to change the axis of its advance and head south against the great industrial city of Stalingrad on the western bank of the Volga River. This was demanding too much of his armies, but despite warnings from his generals that their forces would be overstretched, Hitler insisted his orders be carried out.

The advance forces arrived at the outskirts of Stalingrad in August and more troops and weapons followed. At first it seemed like a relatively easy campaign. With the Germans receiving plenty of supplies and the *Luftwaffe*

Figure 26. Soviet mortars ready for use with the Red Army.

maintaining air superiority, the army was guaranteed support. But first appearances crumbled as Soviet resistance strengthened. As the fighting intensified, so the expenditure in ammunition increased dramatically. By the end of September, after the first full month of fighting, the 6th Army had fired 23 million rounds of small arms ammunition. In addition the artillery and tanks had fired 685,000 shells and the troops had thrown 178,000 hand grenades and fired 750,000 mortar bombs. It was a level of expenditure in ammunition that would increase as the battle spread to encompass the whole of a city which, by the end of December, lay in ruins. As another winter set in, the demand for ammunition and fuel became unsustainable, with German supply lines overburdened and overstretched.

The supply routes were broken completely when the Soviet Army encircled the city in a massive manoeuvre designed to isolate the Sixth Army from the rest of the German forces. There was nothing coming in by road or rail, and in an effort to support the army the *Luftwaffe* tried to fly in supplies. Despite promises of 600 tons of supplies per day, the air force could not meet even the most basic needs of the besieged army. Mortar bombs were lighter than artillery shells and these were transported in preference, but eventually the supply of this ammunition also failed. The Soviet Army on the other hand had no problems with resupply, though production had to be increased to meet demand. Manpower levels were not a problem to the Soviet Army either, whereas the German losses could not be made good. The end came on 30 January when the German commander, Field Marshal Friedrich von Paulus, surrendered. The battle had cost the Germans 300,000 men killed and wounded, with more than 100,000 taken prisoner. Germany's allies had sustained 450,000 casualties. It was the turning point of the war in the east.

At the time of Stalingrad a Soviet infantry division had 9,500 men organized into the standard triple formation with regiments and battalions. The military planners reorganized this structure to distribute weapons to provide a division with an increase in artillery support and a company with six 120mm mortars, which gave each regiment more field guns and 160 mortars. Further changes were made in the structure which, by 1944, saw a Soviet tank corps equipped with a specialist mortar regiment within its organization equipped with twenty-four mortars of 120mm calibre, and each of the three battalions of the mechanized infantry brigade had six 82mm mortars. In April 1945, when the Red Army was fighting in the suburbs of Berlin, even cavalry divisions with a troop strength of 5,040 men, with 5,128 horses and 130 vehicles to tow anti-tank guns, would be equipped with eight heavy mortars and supplied with trucks to transport the division's eighteen medium mortars and forty-eight light mortars, along with the ammunition required for the weapons.

By 1944, a German infantry division fighting on the Eastern Front had 12,352 troops still divided into three regiments, though manpower shortages had reduced the number of battalions down to only two for each of these. By now the light 5cm mortar had been taken out of service, but some units continued to use the weapon as long as they had supplies of ammunition. The situation with manpower levels for the German Army continued to worsen, and by 1945 an average infantry division had barely 7,000 men and was desperately short of weapons, ammunition and other essential supplies such as food, fuel and medical support. By comparison, the Soviet Army infantry division had 9,200 men with three infantry regiments, each with three infantry battalions. Part of the artillery support for the divisional firepower was provided by a specialist mortar company with six 120mm calibre heavy mortars. In contrast, a German panzer division had 13,725 men with an armoured regiment and a motorised regiment of panzer grenadiers, each divided into two battalions and each equipped with four 120mm and six 81mm calibre mortars. A Soviet tank corps had 10,500 men, with an integral regiment of mortars equipped with twenty-four weapons of 120mm calibre, and three battalions of infantry, each with 600 men and their own 81mm mortar units. On paper the German divisional strength was greater but in reality the manpower levels were very rarely operational. Whilst Soviet manpower levels were lower than the German equivalent, they had more divisions deployed overall to give greater troop concentrations. The problems the Germans faced were further compounded by lack of resupply with ammunition and replacement weapons. The Soviets' logistical routes, on the other hand, were secure and supplies of ammunition, weapons and reinforcements could be moved without fear of being attacked.

Chapter Eight

More Operations and Other Theatres

The Japanese attack against the American Pacific fleet at Pearl Harbor opened up a new and very different theatre of operations. Japanese forces expanded across the Pacific and established a perimeter to encompass a radius of some 5 million square miles, mainly consisting of ocean, scattered across which were thousands of islands. The larger islands such as Tarawa, Saipan, Guadalcanal and Vanuatu proved strategically useful and airfields were established on them which were garrisoned by strong protective forces. The smaller islands, such as Makin, required a garrison of fewer troops but were not so strategically valuable. Whilst Japan's success looked impressive it meant that Japanese forces were split and supplies of arms, ammunition and food had to be divided to support these disparate groups. The suddenness and power of the Japanese attacks took the British and Americans by surprise, and for a while it appeared nothing could stand in Japan's way. On 16 December, Japanese troops landed at Victoria Point on the southern tip of Thailand and over the next four months they forced the British back over 1,000 miles to the very border of India. On 25 December, the British garrison of 12,000 troops stationed at Hong Kong surrendered. On 15 February 1942, the British garrison on Singapore, comprising over 138,000 British and Empire troops, surrendered after a fierce fight. The turn of the Americans came on 9 April, when the Philippines garrison of 12,000 US troops and 64,000 Filipino allies surrendered after a week of fighting. During this campaign American troops were still using some old-fashioned types of weapon, including the M1903 Springfield rifle and the 3in Mk1A2 mortar, a slightly improved version of the British weapon designed by Wilfred Stokes during the First World War. Fortunately it could fire 81mm ammunition of the M1 mortar which was in service by this time. Some Mk1A2 mortars not issued to combat units were put to use as training weapons and these remained in service throughout the war.

The losses in troops, equipment and weapons had to be made good, and as the British Army already knew from experience at Dunkirk in 1940, it would take time. New regiments were created, specialist units trained and to equip these men the armaments factories turned out more weapons. Among the new regiments formed were the US Army Rangers, which deployed units to fight in the Pacific and Europe. The 17th Airborne Division was raised in 1942 to serve as a parachute unit. The 101st Airborne Division was established in August 1942, adding to the new form of airborne warfare to which the US had to adapt. The 101st Division had been raised in 1918 but stood down at the end of the First World War. As an airborne division, it parachuted into France and fought across Europe after D-Day. The 82nd Division had been raised in 1917 and fought in France, but after the war it was placed on the Organized Reserved List. In February 1942 it was re-designated, and the following month the division was placed on the Active List and organised as an airborne division. It too parachuted into France on D-Day and fought across Europe. The British Army had begun creating its first airborne units in 1940, having seen the successes of the German *Fallschirmjager* parachute troops at Eben Emael in May 1940. The first units were trained and ready by 1941, and eventually there would be the 1st and 6th Airborne divisions. They would serve across Europe after D-Day. Even though such troops were elite forces, they carried standard infantry weapons and used the same mortars in service. The Italian and Japanese armies also formed parachute regiments but they were never used on the same level or achieved the same successes as their Allied counterparts. The Imperial Japanese Navy raised parachute units, with the basic regiment being known as *Teishin Rentain* (Raiding Regiment) and the airborne group the *Teishin Dan* (Raiding Group). They were armed with the Type 89 'Knee Mortars' and other calibre weapons.

The Allied troops intended for use in European operations were separate from the forces deployed to the Far East, and the generic organisation of these divisions would remain basically unaltered until the end of the war. The numbers of troops increased so that by June 1944 the level of forces held in readiness for D-Day in Britain, for use in subsequent operations and reserves, reached a level of 2,720,000 all ranks. A year later this figure had increased slightly to 2,920,000, from which reinforcements would be allocated to replace losses in units across Europe. The British involvement in the landings at Normandy were the largest operation of the entire war and required ten infantry divisions and two airborne divisions, along with armoured divisions and other specialist support units. On the actual day of the landing, almost 60,000 British and Canadian troops, along with 8,900 vehicles, would be landed on the three beachheads allocated to them. By late July, there were a combined number of 631,000 British and Canadian troops in France, along with some 153,000 vehicles of all types.

New Formations
Infantry training for airborne troops was rigorous and often involved cross-country runs and forced marches in full kit designed to cover a measured

distance in the quickest possible time. Wally Parr, who served as a corporal with D Company the 2nd Battalion Oxfordshire & Buckinghamshire Light Infantry, which was a glider-borne unit, would land in a Horsa glider to capture the bridge across the Caen Canal on the night of 5/6 June 1944. He remembered how, during training, their commanding officer Major John Howard would take his men out twice a month for route marches, during which they would cover a distance of 22 miles in 5½ hours, carrying kit and weapons including rifles, Bren guns and mortars. This was an incredible feat of endurance, especially when one considers that today in peacetime conditions some athletes, wearing only singlets and shorts and running shoes, complete marathons of just over 26 miles in the same time. Airborne forces were elite and the training regime was designed to be tough, not only to build up physical fitness but also to filter out those who could not meet the exacting demands. Hardwick Hall at Chesterfield in Derbyshire was one such training centre, where the regime was physically and mentally demanding and included running 200 yards in 16 seconds wearing full kit. Other exercises called for recruits to complete a forced march of 8 miles in 75 minutes. Airborne forces in other countries, including the German *Fallschirmjager*, were trained equally hard and taught how to handle a range of weapons, including those used by the enemy, under all conditions including night operations. Veteran Bill Guarnere, who served with the 2nd Battalion, 506th Parachute Infantry Regiment (PIR) of the 101st Airborne Division, remembered how the training at Camp Toccoa in Georgia was 'brutal – all physical conditioning'. The 3/506th PIR showed what it was capable of by covering a distance of 136 miles from Atlanta to Fort Benning in 72 hours carrying full equipment and weapons in foul weather. It could be argued that such a comparison is not balanced, but it does say a great deal for the levels of physical fitness of the fighting soldier in 1944.

A British airborne rifle section was made up of nine men, including a machine gun group with three men to serve the Bren gun. This group also carried the 2in mortar which was used to fire mainly smoke bombs in this role to screen movements. The other six men in the section were armed with either the .303in calibre bolt-action rifle or the 9mm calibre Sten sub-machine gun, and each man would also carry extra ammunition for the mortar. By the time of the Tunisian campaign, conducted between November 1942 and May 1943, British airborne units were using the 3in mortar at battalion level to fire HE bombs to provide support. The 2nd Parachute Regiment issued Standing Orders which told crews operating 3in mortars in the campaign that 'alternative positions [be] prepared in addition from which they can bring fire down on their own positions in the event of being overrun.' Experiences learned in North Africa would serve the British airborne troops well in Europe, where the Germans used their mortars with great skill to exact a terrible toll. During one stage in the campaign, Lieutenant Colonel John Frost voiced his impatience at the inability of his lightly-equipped forces to engage the German machine gunners because they were 'not within the range

Figure 27. British Airborne paratrooper carrying a 2in mortar in his pack.

of our Brens'. The Bren gun had an effective range of 600 yards, firing 500 rounds per minute from box magazines. The Germans used the MG42, which could fire 1,200 rounds per minute at ranges of 2,000 yards or more. Lieutenant Colonel Frost also lacked 3in mortars which could have been used against the enemy machine guns, the fire support role they were designed for.

By the time of Operation Torch, the Anglo-American landings in North Africa on 8 November 1942, there had been attempts to test German defences along the French coast. The largest such operation, codenamed 'Jubilee', was carried out on 19 August that year. It was a costly disaster in terms of men lost, along with equipment and weapons. However, the Allies learned valuable lessons from these mistakes, which would help in the planning for D-Day as they finally opened up the second front which Joseph Stalin, the Soviet leader, had been calling for. The Allies' success in concluding operations in North Africa in early 1943 freed up resources for the invasion of Europe. Meanwhile, the Allies decided to keep up the pressure on the Germans and Italians in the Mediterranean and an amphibious assault by 180,000 troops was made against Sicily. Fierce fighting by infantry across the island saw the use of mortars. Jack Swaab was an officer serving with a field regiment of Royal Artillery. He had already seen service in North Africa and was called into action again on Sicily in July 1943. In his 2005 book *Field of Fire*, which is his war diary, he remembers being fired on by German mortars and soldiers being wounded by splinters from mortar bombs. In his recollections he wrote how at the Paterno-Catania sector he believed the Germans were 'concentrating for one last stand' and recalled the 'sinister crump of mortars which stopped movement along tracks'. He later fought in mainland Italy and then in north-west Europe, landing in the Sword Beach sector on D-Day, 6 June 1944. He went on to serve throughout the campaign and eventually ended the war in Germany in March 1945. During the campaign he maintained a healthy respect for the effectiveness and accuracy of German mortar fire. Even in late March 1945, he reports heavy barrages by mortars which continued to cause a great many casualties right until the last days of fighting.

The last of the fighting on Sicily was concluded on 17 September, by which time the Italians had managed to evacuate 62,000 troops to the Italian mainland. The Germans, meanwhile, evacuated 40,000 troops along with much of their weapons and equipment. The operations on Sicily led to the deposing of Mussolini as the leader and split the loyalties of Italian troops. During the long, hard-fought campaign in Italy, the Allies mounted two further amphibious landings, at Salerno on 9 September 1943, six days after the Allies had first landed in Italy, and Anzio in January 1944. The latter was an attempt to outflank the Germans and get behind their position.

Italy proved to be a harsh battleground, with every mile having to be fought for. The Germans constructed a series of fortified defensive lines at intervals such as the Caesar Line, Hitler Line and the most formidable, the Gothic and Gustav Lines. Into these defences hundreds of mortars were emplaced. The Germans developed the tactic of siting their mortars on the reverse side of slopes so as not to be immediately spotted by the Allied observers looking and listening for the tell-tale signature of the weapons being fired. Because mortars fire indirectly, as long as the Germans had a forward observer to correct their own fall of shot, the mortar batteries in such positions could fire with relative impunity and could only be attacked properly by aircraft or artillery barrage.

In the Gothic Line, the Germans built 2,000 defensive positions. The 1st Battalion, London Irish Rifles were shelled so heavily by artillery and mortars they referred to their location opposite the Gothic Line as 'Stonk Corner'. The US 91st Infantry Division attacked part of the Gothic Line on 11 September and came under intense mortar fire. During the fighting to penetrate the defences, General Livesay directed that troops 'fire all the ammunition you can haul'. Between 11 and 22 September, some 168 pieces of artillery attached to a division fired in excess of 94,000 rounds. At other locations mortars were causing great consternation among the Allies, and Major Desmond Woods, MC, serving as a company commander with the 2nd London Irish Rifles, recalled how his troops approaching the positions at Monte Cassino received: 'a certain amount of trouble from their [Germans] mortars'. He continued: 'they were able to put their mortar shells down and this mortaring could become very unpleasant and I did have a certain number of casualties merely from mortar fire.' Artillery was the heavyweight but mortars, being lighter, were more mobile and easier to relocate to a new firing position, and it was the weight of fire they could put down on the Allied positions which caused the most problems. The Germans holding the high ground at Monte Cassino were able to make good use of the rubble of what had once been the magnificent abbey after it was bombed by the Allies, and were able to take shelter among the debris. Among the defenders holding the site were elite *Fallschirmjagers* who used all means available to them, including extensive use of mortars. The fighting for Monte Cassino was costly and the advance up the length of Italy was ponderous. All the while the Germans and some stalwart Italian troops fought well, even though they were short of supplies. The Allies, on the other hand, were well, supplied with ammunition, weapons and troops. They maintained the pressure, but the last German troops would not surrender in Italy until 2 May 1945.

Figure 28. German 8cm GrW34 in action.

At the same time that the Allies were meeting stubborn German resistance in Italy, the Soviet Red Army was massing for a huge offensive on the Eastern Front. The Red Army had relieved Stalingrad six months earlier and now it was keeping up the pressure. Both sides began amassing their forces for an engagement around the city of Kursk in between the Central Front and the Voronezh Front. The fighting lasted from 5 July to 23 August 1943 and eventually developed to extend along a frontage of 1,200 miles. At first it seemed as though the Germans might just pull off another victory, but eventually the Soviet weight of forces turned the battle in their favour. The Germans assembled some 912,000 troops, with almost 3,000 tanks and armoured vehicles and fewer than 10,000 pieces of artillery and mortars. Facing them was a Soviet force which would increase in strength to a level of 1.9 million troops, over 5,000 tanks and other armoured vehicles and 25,000 pieces of artillery and mortars. The exact numbers of mortars used in this battle is difficult to judge, but levels were high in all calibres up to and including the very heavy calibres used with batteries of conventional artillery. After several weeks of fighting, the Red Army emerged the victor. The German Army was smashed, yielding prisoners and captured stocks of weapons. With the Allied success on Sicily, the German Army was now under severe pressure on two main fronts.

D-Day and the Second Front

In England military commanders were making plans to co-ordinate the invasion of Europe. More American troops were being trained and assembled for the campaign which lay ahead and more weapons and specialist equipment was prepared. Rigorous training for amphibious operations was undertaken for assaulting from specialist vessels called landing craft. These were Higgins boats, named after designer Andrew Higgins, and were 36ft in length and almost 11ft in beam (width). They were capable of carrying up to thirty-six troops. For the Normandy landings, thousands of Landing Craft Vehicle/Personnel (LCVP) would be used to ferry troops, vehicles and heavy weapons ashore. To land the infantry as quickly as possible these craft were fitted with ramps in the bow which could be lowered to allow the troops to deploy directly onto the beach landing zones. The Americans designated some Higgins boats as 'Assault Boats' along with others known as 'Support Boats', each of which carried a compliment of twenty-four men organised into specialist teams along with their equipment. The Assault Boat Team landing craft ferried a team equipped to cut the barbed wire entanglements, riflemen, an anti-tank team armed with a Bazooka and a flamethrower team. It also carried a four-man mortar team to operate the 60mm M2 mortar. The No. 1 was the observer who carried an M1 carbine as his personal weapon along with compass, binoculars, cleaning kit, torch and twelve rounds ready to use. The No. 2 was the actual firer of the weapon who carried five rounds ready to use and a pistol for personal protection. The No. 3 was the assistant firer or gunner who carried twelve rounds and an M1 carbine. The No. 4 was the ammunition carrier who carried a further twelve rounds and an M1 carbine to give the crew a total of forty-one rounds ready to fire. The Support Boat landing craft also carried specialist crews such as a demolition team to deal with obstacles, flamethrower team and anti-tank team with a Bazooka. This boat carried an eight-man team to operate and protect the M1 81mm mortar. The No. 1 was the observer and carried binoculars, compass, five rounds ready to use and an M1 carbine. The No2 carried the bipod and an M1 carbine whilst the No. 3 carried the barrel and aiming stakes along with an M1 carbine. The No. 4 carried the baseplate and an M1 carbine. The No. 5 carried seven rounds ready to use, 400 yards of telephone cable for communications and his personal weapon, which was an M1 carbine. The three remaining men in the crew were each armed with M1 carbines to protect the mortar crew and each of them also carried seven rounds of ammunition, giving a total of thirty-three rounds ready to use once they had landed and deployed.

Plans for D-Day had been laid and changed many times as amendments were made and information concerning the disposition of German troops built a clearer picture. The Normandy coastline was finally selected for the landing because of the beach areas and the fact that the Germans, who had always believed that the landings would be made in the Pas de Calais region, would not be expecting the assault to come in this area. A fifty-mile stretch of the coastline was divided into five beachheads, with airborne divisions landing on the flanking edges to protect these vulnerable zones. The two British beaches were codenamed Sword and Gold, the Canadian beach was Juno and the two landing beaches designated for American landings were codenamed Omaha and Utah. For many troops landing on 6 June, this was to be their first real exposure to enemy fire. They quickly discovered it was unlike anything they had experienced during live firing exercises at places such as Woolocombe in North Devon. In the confusion of battle there was the inevitable disorientation as some troops landed in the wrong place and others became separated from their units, making them targets for artillery and mortars. According to the training manual of the US Army, on landing the mortar team was to 'move to a position where it can bring fire on possible targets with maximum protection; they will receive orders from the assistant section leader to bring fire on open emplacements, but may direct fire on such emplacements at his own initiative. Fire is directed until the successful conclusion of the assault, when the mortar is prepared to fire on enemy concentrations and to break up enemy counterattacks.' Written in a drill manual and performed during training exercises, such information sounded simple. It was Carl von Clausewitz who, in his nineteenth-century book *Vom Krieg* (On War), stated that 'War is simple, but in war the simplest things become very difficult.' So it was when the mortar teams landed on the beaches at Normandy. Chaos was everywhere due to the intensity of enemy fire. On Omaha Beach, for example, things began to look particularly bad as plans collapsed and casualties increased, with support weapons and their crews being lost. Eventually the troops were organised, officers regained command of units and orders were passed. The potential disaster on Omaha was averted and the assault managed to move forward and off the beach.

The Germans had been expecting an invasion, but where and when were unknown factors. Some officers, like General Erich Marcks, commanding LXXXIV Corps in Normandy, guessed that the landings would come in the Normandy sector, but although he stressed his belief, he was either dismissed or ignored. He held the opinion, as did his commanding officer, Field Marshal Erwin Rommel, that any Allied landing must be stopped on the beaches. He stated: 'If you don't succeed in throwing the British back into the sea, we have lost the war.' In 1943, Rommel had predicted that: 'The west is the place that matters. If we once manage to throw the British and Americans back into the sea, it will be a long time before they return'. Marcks, like Rommel, was a professional soldier, having served in the First World War and for the Weimar Republic in the years between the wars. He knew what he was talking about. He had also served in Poland in 1939, during the French campaign in 1940 and in 1941 had lost a leg due to severe wounding whilst fighting on the Eastern Front. With such battle experience his warnings should have been heeded. However, the overall commander, Field Marshal Gerd Von Rundstedt, wanted to allow the Allies to come ashore and then use the panzer divisions to destroy them completely. Unfortunately the man-power shortage meant the troops were low grade, armour was deployed else-where and resupplies of fuel and ammunition were often unavailable. Marcks was killed when an aircraft attacked his car on 12 June. A 20mm cannon shell hit an artery in his remaining leg and he bled to death by the side of the road. Defensive preparations had been made and in some cases the defences were quite formidable. Rommel had done his best with the time and resources available to him, but the Allies attacked before he could complete his plans.

The landings over the beachheads at Normandy were planned to go in at specific times, which differed to take into account the tidal conditions. During the actual landings, mortars were essential and truly became the 'infantry-man's artillery'. As the troops assaulted across the sands these were the only weapons which allowed them to hit back with something other than just their rifles. Rifle grenades and hand grenades were limited in range and light mortars – carried at platoon level – had a greater range than either. On Omaha Beach the landing area was overlooked by high ground which rose to 100ft in places. The German defenders holding this ground could select targets and engage them with machine guns and mortars. The Allies had air support and naval gunnery, but the man on the beach in amidst the melee of battle wanted something with which he could hit the enemy. This was the mortar and in particular the 60mm M1, which could be carried by the crew and brought into action very quickly without need for great preparation. In the rush to get ashore and taking cover from enemy fire, much equipment and weaponry was dropped, including mortars, which were essential for local fire support. Lieutenant John Spaulding lost his rifle as he landed on Omaha Beach as part of 1st Division and remembered that his group did not lose any men in the deep water because they helped one another, and could only do so by abandoning equipment. At the time it seemed the right thing to do but they would come to regret it later. Lieutenant Spaulding's men had lost all their heavy weapons during the wade ashore and 'we lost our mortar, most of the mortar ammunition'. They also lost other vital equipment which would have allowed them to tackle obstacles and pillboxes. Without this equipment and their mortars to give them fire support, they were left with no choice but to wait for armoured support before taking on bunkers and other gun emplacements.

The fire from German mortars on the high ground looking down onto the landing site designated Omaha simply had to find their range. Accuracy did not have to be particularly good as the mortar bombs landed indiscriminately and 'worked over' the ground, dismembering even further those already dead and killing and wounding fresh victims. Captain Mabry was one of those rushing ashore to find shelter amid all the chaos. He remembered the man in front of him being caught in the blast of a mortar bomb. He felt a sharp blow to his abdomen and thought he had been wounded by the same blast. Looking down and fearing the worst, he saw to his relief that instead of iron from the mortar bomb he had been hit by a thumb from the poor unfortunate he had witnessed being killed. Indeed, the effectiveness of German mortars firing onto the American landings on Omaha Beach is mentioned in many memoirs written by veterans, such as those of Jack Womer, who served with the 29th Ranger Battalion before transferring to the 101st Airborne Division. In his book *Fighting with the Filthy Thirteen* he makes many references to how the Germans deployed mortars and machine guns and sited them together to cover avenues of approach along paths and roads. As infantry advanced, the two types of weapons would open fire and inflict heavy casualties. The US assessment of 80 per cent casualties caused by mortars is supported to a degree by British figures, which propose a slightly lower total of around 75 per cent of casualties being caused by mortars, grenades, aerial bombs and shells, with 10 per cent being caused by small arms fire and anti-tank shells and the remaining 15 per cent due to miscellaneous causes such as mines, blast and crushing. The same results were being recorded in other armies, including those of the Germans and Soviets. During the campaign in Tunisia, the German 8.1cm mortar had been credited with causing around 40 per cent of all shell shock cases reported, such was the intensity of fire an experienced crew could lay down on target areas.

British troops began to land on Gold Beach at 7.20am amid 'murderous fire', according to Sergeant G.E. Hughes serving with the 1st Battalion, Hampshire Regiment, who landed at Arromanches. His unit went on to clear

Figure 29. German GrW34 mortar on its bipod, the weapon that caused many casualties during the Normandy campaign.

three villages and over the next six days of continuous fighting he recorded in his diary that German mortar fire was 'undescribable [*sic*]'.

Able Seaman Ken Oakley, serving with F Commando, landed in the area of the coastal town of Lion-sur-Mer, in the sector given the code name Queen Red on the western end of Sword Beach. He remembered running ashore from his landing craft along with others as 'mortar and machine-gun fire sped us on our way'. Advancing further up the beach, they found some shelter from the effects of direct machine-gun fire, but the indirect fire from mortars was creeping towards their positions. Oakley remembered how: 'It seemed as if we were just outside the mortar-fire pattern and suddenly a DD [Duplex Drive or Swimming] tank loomed up behind us. The hatch opened and a voice called out, "Where is the fire coming from?" I answered, "A couple of hundred yards to the right at 45degs". "OK," he said. The hatch banged shut, then Bang! The shell screamed over our heads. It was no contest, the mortar fire ceased, the machine-gun fire subsided and just the occasional sniper shot rang out.' Incidents like this were being experienced all along the 50-mile wide beachhead made up by the five landing beaches. Tanks and other specialist support vehicles included the Centaur, which was a Cromwell tank armed with a 95mm calibre gun for close support roles and to destroy targets such as gun positions. Private Richard Harris landed on Sword Beach with the 1st Suffolk Regiment and remembered the occasion as being 'very confused' and how 'between the crunch of mortar bombs and the whizzing sounds of shells and chunks of metal flying about', men still went about their tasks.

Some members of the 1st Infantry Division (the Big Red 1) had fought in North Africa and the veterans from that operation knew what to expect, but those uninitiated troops landing on Omaha learned very quickly to fear and respect concentrated fire coming from the well-sited German mortars firing onto the beach. Wesley Ross, who served with the 146th Engineer Combat Battalion which landed on Omaha Beach, remembered how as his landing craft approached the shore 'small arms fire and artillery and mortar fire began kicking up spray around us'. The Germans had had time to prepare their positions and fired from overlapping strongpoints built from concrete and known as *Widerstandsnest*, which were all given a tactical WN number reference. On Omaha there were five natural routes leading inland from the beach, which were ravines known as draws which served as obvious exits from the landing point. At Hamel-au-Pretre near Vierville-sur-Mer, the Germans had installed a range of defences looking down on the beaches. The site designated WN70 included an 8cm mortar in its weaponry and from its position it was able to fire down on the American troops coming ashore at the landing point designated as Dog Green. American troops bunching up at these points

Figure 30. German troops assemble a GrW34 ready for use.

seeking shelter became targets for mortar barrages and machine-gun fire. It has been estimated that at least 75 per cent of all casualties on Omaha Beach were caused by mortar bombs. It is a figure shared by the late historian Stephen E. Ambrose in his book *The Victors*. Another historian to support this figure is Anthony Beevor, who in his book *D-Day The Battle for Normandy* states that: 'Three times as many wounds and deaths were caused by mortars as by rifle or machine-gun fire.' A detailed technical analysis of German weapons in the aftermath of the Normandy campaign concluded that the German 8cm GrW34 mortar inflicted a casualty rate of 65 per cent for every fifty-seven bombs fired, which made it arguably the most effective weapon of the campaign and indeed far more efficient than machine guns in terms of ammunition expended to inflict casualties.

John Martino, serving with the 81st Chemical Mortar Battalion attached to the 116th Infantry Regiment, landed on Omaha Beach and remembered how 'the beach was under heavy mortar fire'. The troops landing could rely on fire support provided by the naval task force to destroy artillery positions and other strongpoints; the 1,630-ton destroyer USS *Frankford* provided fire support using her four 5in guns to assist troops pinned down by machine guns and mortar fire. The ship's Action Report for the incident recalls: 'After close observation the exact location of the battery was noted at 10.32. At 10.36 commenced firing on the battery using direct fire, range about 1,200 yards. On the fifth salvo a direct hit was obtained, a large cloud of green smoke was noted and the mortar battery ceased firing. Our troops then advanced and a number of German troops were seen to surrender.'

As they pushed inland, the Americans thought their casualty rates might fall, but they soon discovered that the fighting was not to be all one-sided in their favour as they moved into the area of the countryside known as the *Bocage*. Casualty rates began to increase again as mortars firing in support roles slowed down the advance. The Normandy countryside is farming land with the fields edged by extremely thick hedgerows – the *Bocage*. The openings in the hedges drew the advancing troops towards them like magnets, but the Germans had these features covered with mortars and machine guns. The Germans were masterful at regrouping and mounting stiff resistance, and referred to the fighting in the *Bocage* as *schmutziger Buschkrieg* (dirty bush war). They knew how to use the terrain to their advantage. They would often hold fire and observe the Allied infantry advancing behind cover of tanks and aim their weapons carefully. Then, with unerring accuracy, 8cm mortars would open fire, causing the troops to scatter and seek cover. As they were running the machine guns would open fire to exact a terrible toll on the exposed infantrymen. Even before the last rounds had landed, a well-trained and experienced mortar crew would have packed up and left the firing position and moved to a new location before counter-battery artillery or an airstrike could be called in to obliterate the old position.

Similarly, openings in entanglements of barbed wire and paths through minefields were also covered by a combination of these weapons and others. Despite their training, Allied troops unwittingly entered these killing zones, which were fired on by pre-sited weapons. The Germans knew the range coordinates, which were recorded on their maps. All a crew had to do was point the barrel of their mortar in the right direction, elevate it to the correct angle and begin to fire. They did not even have to fire a ranging shot for correction. It was sudden and unexpected. Firing could stop as quickly as it had begun as the Germans moved their positions to another location. This tactic was repeated time and time again and added to the mounting number of casualties. The British discovered how German mortars could open fire and inflict casualties with startling effect. Brigadier J.O. Cunningham of 9 Infantry Brigade was conferring with officers on 6 June when several mortar bombs landed, killing six officers and wounding six more, including Cunningham, who was unable to pass on orders. The incident was over in an instant and the mortar crew would have moved to a new location before any action could have been taken against them. A similar incident befell an armoured unit from the Sherwood Ranger Yeomanry during the Normandy campaign when a mortar bomb landed in the midst of a group of men discussing orders. Lieutenant Stuart Hills was walking over to the main group when the German mortars opened fire. He remembered joining the group just as a bomb landed, but was not hurt by the explosion. A sergeant sitting by his side was slightly injured by the blast. When he returned to his tank, Hills discovered one of his crew had been killed so suddenly by the mortar fire that he still had a cigarette in his mouth.

One US Army veteran of this type of fighting remembered: 'When our men appeared, laboriously working their way forward, the Germans could knock off the first one or two, cause the others to duck down behind the bank, and then call for his own mortar support. The German mortars were very, very efficient. By the time our men were ready to go after him, the German and his men and his guns had obligingly retired to the next stop. If our men had rushed him instead of ducking behind the bank, his machine gun or machine pistol would knock a number off. It was what you might call in baseball parlance, a fielder's choice. No man was enthusiastic about it. But back in the dugout [rear area] I have often heard the remark in tones of contempt and anger; "Why don't they get up and go?".'

It was all too easy to criticise from the safety of the rear zone and the experienced Germans knew how to use the combined firepower of mortars and machine guns to maximum effect. The advancing American infantrymen, many of them still relatively inexperienced in combat, were learning as they went forward and had every reason to be cautious. The initial landings on Omaha had cost them more than 3,800 killed and wounded out of the force of 34,250 landed on 6 June. At Utah, where 23,250 men were landed, the fighting was not as intense and the casualty rate was very light, with only 210 killed and wounded. The German defences at this landing point were designated as WN5 and included a comprehensive range of light and heavy weapons, including anti-tank guns. At least one emplacement was armed with an 8cm mortar, which was destroyed by naval guns almost as soon as it opened fire

The parachute troops had an equally tough time being fired on by mortars. Private Carwood Lipton, serving with the 506th Parachute Infantry Regiment of the 101st Airborne Division, became one such casualty during a mortar barrage when a bomb landed several feet in front of him. He received multiple wounds from splinters, including to his left cheek, an artery in his right wrist was punctured and his right leg in the area of the groin, where arteries and tendons are located, was also hit. His wounds were tended by a medic, who put a tourniquet on his wrist to stem the bleeding. His groin wound was checked and he was given a shot of morphine for the pain, taken to a field hospital and later evacuated back to England. Incredibly, Carwood survived these wounds and later returned to service, seeing further action with his unit during the Battle of the Bulge in late 1944, and ended the war holding the rank of lieutenant.

The British airborne elements of the invasion had landed during the night of 5/6 June to the eastern edge of the beachhead area, and some glider-borne units also landed further inland. These units were given a series of tactical tasks which included seizing bridges and crossroads and neutralising artillery positions, such as the battery of guns set in concrete bunkers at Merville which could fire on Sword Beach. Theses airborne elements came under mortar fire as they consolidated or attacked a position. Another vital point to be captured was the bridge across the Caen Canal and river. Three gliders were given the task of securing the bridge, which was achieved with almost total surprise in fifteen minutes. As the gliders touched down, the men of the Oxfordshire and Buckinghamshire Light Infantry, in the role of airborne landing troops, secured the area on both sides of the river. The Germans responded by using their machine guns and mortars to engage the British troops as they fanned out to take up positions. Staff Sergeant Potts of the Glider Pilot Regiment, who took part in this operation, recalled how he believed the mortar was the 'one great dread' among the troops because they could hear the bombs going over their heads with what he described as a 'wobbling' sound. He said they could hear this wobbling noise and instinctively ducked down to take cover and hoped for the best.

Nearly all troops remember how mortars made a particular sound on being fired – a loud 'popping' noise. For example, Ron Davison, who served with the East Surrey Regiment, recognised mortars being fired as making a 'plop' sound after which 'the bombs came in'. Sergeant 'Paddy' Jenkins, serving with 3 Parachute Regiment during the Normandy campaign, recalled the sound of mortars firing as being a 'thwump'. *SS-Hauptsturmfuhrer* (Captain) Hans Moeller, serving at Arnhem during Operation Market Garden, recalls 'muffled "dumpfs" signalled the barking of mortars'. Jack Burke, serving with A company of the 5th Rangers, landing in Normandy, remembered that: 'Artillery shells have a whistling sound, you only hear it maybe two seconds before it explodes. It is a definite sound that anyone in combat recognises, and the general effect is to yell "get down" or something similar. Incoming mortar shells have a "whooshing" and you don't have much time to hit the dirt, possibly a second. If the mortar is firing close you can hear a "thump" as the shell is dropped down the tube, but that is rare.' When the sound of mortars being fired was identified, whichever way the noise was described, everybody took cover.

Lieutenant Dennis Rendell recalled hearing the 'whine of mortar bombs falling very close'. He continued how, after the bomb had exploded, his: 'sole recollection is a ghastly smell of sulphur'. Germans who fought at Stalingrad in the winter of 1942 remembered the 'whir and whistle' of Soviet mortars bombs. Some men recalled the sound of exploding mortar bombs as detonating with a 'crump' but the difference in the sound could be explained by the type of ground the bombs landed on. Major Bill Apsey, who served with the 11th Armoured Division in Normandy and was badly wounded by a mortar bomb, held a much different opinion. He said 'Damned mortars – you could never hear them coming'. French roads were often surfaced with hard cobblestones and when mortar bombs fired from whichever side landed on these the resulting blast was multiplied. Splinters of metal were thrown out in a 360-degree radius, mixed with shards of stone to produce lethal wounds. Lieutenant Colonel Frost remembered this effect from his time during the Tunisian campaign and recalled how 'their [the Germans] mortar fire was all the more deadly on the rocky ground'. On the landing beaches this was not so much of a problem – a lot of the blast would be absorbed by the soft sand. Nevertheless, the blast would still cause concussion and even death through the pressure of the blast wave. During the fighting across Normandy, German defenders would often hold out in villages which had to be cleared out each in turn. In the see-saw of battle, some of these towns and villages changed hands several times or more. Some became the centre for intense fighting. For example, Tilly-sur-Seulles had 70 per cent its buildings destroyed before the area was finally secured by Allied forces.

The ground over which they were fighting was well known to the Germans, especially the troops of the 352nd Infantry Regiment, who had trained for such an operation. They were able to deploy their mortars to pre-sited positions from where they could fire on targets. Initial German reaction to the invasion was relatively slow, but gradually over the coming days they moved more troops and armour into areas identified as the axes of advance. On 10 June, the German 3rd *Fallschirmjager* Division began to arrive after a four-day journey by truck from Brittany. They had been forced to drive by night because Allied air superiority made it too dangerous to move on roads by day. This well-equipped division had almost 16,000 men, and although most had never been in combat before, like their American opponents, they were well trained and physically fit. Facing them were men of the 29th Infantry Division armed with rifles, machine guns and 60mm calibre mortars. The Germans

Figure 31. The *Kurze* short-barrelled version of the GrW34, as used by the *Fallschirmjager*.

had numbers of machines guns and three times as many mortars as the Americans, and in heavier calibres. They may have had the potential of greater firepower but they could not guarantee a reliable resupply route for ammunition, and replacements for troops lost in battle were few in coming forward. It would be a war of attrition which would be won by the side which had the best and most resources. In this respect it was the Allies who had the upper hand.

Gradually, the Germans were forced back and the Allies, with their superior logistics to replenish ammunition and replacement troops, kept up the pressure. Even as the Germans were withdrawing and not receiving the proper levels of supplies, they quickly developed a new spoiling tactic designed to undermine morale in the most simple, yet effective way possible. Forward observers would watch as the Americans moved into an area and established the position before nightfall. Hot food and drink would be prepared and as darkness descended the German artillery and mortars would open fire on the position. It was totally demoralising and denied the troops proper rest. The Germans would place their mortars in positions approximately 550 to 875 yards in front of the Allies' positions, which made it difficult for counter-battery fire from either mortars or artillery to engage the Germans without risking the possibility of hitting their own troops. This tactic was familiar to Lieutenant Colonel A.E.C. Bredin, who mentions it in his book *Three Assault Landings*, which tells the history of the 1st Battalion, Dorset Regiment, which was engaged in amphibious assaults on Sicily, Italy and at D-Day. Lieutenant Colonel Bredin recalled: 'The German [tends] to lie low during the day; but at the same time maintain the most effective watch on our movements. Any carelessness in exposure or movement on our part, by groups of either men or vehicles, drew the inevitable mortar fire'. It was a problem which would lead to the development of specialist battlefield radar to locate German mortars and allow accurate return fire to destroy their positions.

Arnhem and Operation Market Garden
Three months after the D-Day landings, the Allies launched a daring airborne operation, codenamed Market Garden, to land parachute troops in drop zones from where they could seize a series of vital bridges that would allow Allied forces to advance into Holland and then deep into German-held territory. It was General Montgomery's plan and he believed it would shorten the war by months, driving a wedge into the German Army. The plan involved the American 101st and 82nd Airborne Divisions and the British 1st Airborne Division to spearhead the attack and capture and hold the bridges until the main ground units, comprising British XXX Corps, could link up and secure the route. The first wave of some 20,000 paratroopers were dropped into the three separate zones on 17 September and some initial success was made. Thereafter things slowed down. The American airborne units, after much fighting, seized their bridges and held them to allow XXX Corps to advance. At the Son Bridge, the 2nd and 3rd Battalions, 502nd PIR of the

101st Airborne Division encountered heavy fighting against German forces. Corporal Pete Santini recalled how the fighting developed with it all going 'quiet for a moment and then it seemed like all "Hell broke loose". Mortars, 88mm, 20mm, machine pistols, rifles, and our own mortars, machine guns, Tommy guns, rifles, carbines, shots from all directions, a bedlam of noise which was impossible to describe. This went on all morning.' This firefight was an all-arms engagement as the infantrymen used rifles and machine guns with the mortars providing fire support as the German attackers tried to remove the American defenders from their positions holding the bridge. At Arnhem, the British found themselves cut off and surrounded by superior numbers of German troops equipped with heavier weapons and armour. They were supplied by air drop but this mostly fell into German-held areas. The British airborne force did its best and fought against overwhelming odds, until finally on the night of 24/25 September the survivors were withdrawn from Arnhem.

British Air Landing Brigades which were flown into battle by gliders were armed with the standard infantry weapons, including rifles and Bren guns, and also carried 2in mortars. This form of deploying troops into battle had already been proven in battle by the Germans during their invasion of Holland in 1940 and the British during the assault on Sicily during Operation Husky in July 1943. During Market Garden, British Horsa gliders were used to fly in six 6-pdr anti-tank guns, ten 3in mortars, Jeeps and supplies of ammunition and the crews for these weapons. The American airborne divisions also used their gliders in the same way, and Donald Burgett, who served with the 101st Airborne Division, recalled how, when flying into action in gliders: 'The men with the machine guns, mortars and other heavy or cumbersome gear sat closest to the exit door so they could get out first.' Presumably this was so that they could begin to lay down covering fire whilst the rest of the troops deployed to take up firing positions. Peter Clarke of the Glider Pilot Regiment was one of the first to fly to Arnhem, arriving in a Horsa glider carrying 23 Mortar Platoon of the 1st Battalion, Border Regiment on 17 September. He remembered there were thirteen men in his glider, along with six handcarts to allow the mortars and the ammunition to be transported quickly and easily. On the day that Peter Clarke flew in, the War Diary of No. 2 Wing the Glider Pilot Regiment (later to become the Army Air Corps) records how the Germans reacted to the operation:

17.9.44. Flak very light and landing successful.

18.9.44. Landings were good. Flak much heavier and number of aircraft shot down. Fighting intensified around the defensive perimeter set up by the British airborne troops as the Germans moved in heavier equipment including self-propelled guns and mortars.

This is mentioned in the War Diary at 3.30am on the morning of 21 September, which records: 'A day of continual attack in our sector. The enemy brought down an SP gun into the wood at approx. 690788. In the afternoon 'F' Sqdn counter-attacked the infiltrating Germans in the wood and forced them back. Mortaring was severe and we had considerable casualties.'

The Diary continues:

22.9.44. A message from the Brigadier: 'We're up agin it'. Heavy mortaring and shelling throughout the day but our posns. were held intact. Continued casualties caused our line to become dangerously thin.

23.9.44. Intense shelling and mortaring. Stocks of food and water very low. Many casualties.

24.9.44. In the afternoon a combination of shelling and mortaring, SP guns and a flame-throwing tank forced us to abandon the wood. The survivors took up posns. in the houses across the rd. covering the exits from the wood. No reinforcements could be obtained.

The reference to lack of supplies was not a problem singular to this position. All across the battleground of Arnhem supplies were falling into German hands. The British troops holding the Hartenstein Hotel as their unit headquarters remember coming under fire from mortars which were accurate and the weight of the barrages indicated to them that the Germans were not short of ammunition. Major General Roy Urquhart, commanding the British forces at Arnhem, remembered how he 'could hear the plop and whine of mortars and some of these bombs were falling with unsettling accuracy on the crossroads and in the woodland where many 3rd Battalion men were under cover'. The German mortars were about 2,000 yards distant and some of the men sustained wounds from mortar bombs as they exploded among the branches of the trees, creating air bursts against which they had little or no protection. Major General Urquhart recalled how when the mortar bombs exploded in such a way among the branches of the trees, the shell splinters flew further in all directions than when they hit the ground. Major Digby Tatham-Warter, holding positions at the Arnhem bridge with Lieutenant Colonel Frost, was seen to be taking shelter from mortar bomb splinters by standing under an umbrella, not that it would have protected him, but the image made for good humour to distract the troops in a worsening situation.

On 18 September, the day after the British had seized the north end of the bridge at Arnhem, the Germans mounted an attack spearheaded by armoured vehicles. The British responded with concentrated fire, including PIATs, and they managed to halt the attack and the withdrawing Germans left burning vehicles to litter the bridge. Mortar fire was directed towards the British from across the river and this was directed with precise accuracy. Sometimes the shelling was just too accurate to be down to good map reading. Amidst all the confusion it was noticed how the horn in one of the vehicles abandoned on the bridge in the aftermath of the abortive attack began to sound as though signalling to German positions. After receiving two or three mortar barrages, it was noticed that the vehicle's horn sounded too frequently and distinctively to be just a coincidence. It was noted that whenever the mortar bombs landed on target, the horn sounded more regularly. A rifleman was sent out to investigate and as he made his way forward he noticed a figure in a vehicle. He took aim and fired, whereupon the man in the lorry fell forward, dead. The mortar fire continued but with reduced levels of accuracy. It was finally stopped when British 75mm howitzers fired on the positions. The Germans brought forward 88mm and 105mm guns to replace the mortars, and these did more damage until they too were engaged by the British 75mm howitzers. It had been a strange episode in a desperate action which was to end in disaster for the British troops holding on for a relief that would never come.

The isolated British troops were soon running short of ammunition. They and could not maintain a sustained response with their mortars and were forced to conserve their ammunition supplies. That was hardly surprising considering the Germans held all the surrounding area and controlled all the roads along which their supplies would be carried. They were able to stop the British troops from retrieving the canisters of supplies dropped by parachute. The parachutes were colour-coded to denote the type of supplies: red for weapons and ammunitions; blue for food; yellow for medical supplies; and white for miscellaneous stores such as batteries and items of clothing. All too frequently these canisters landed in German-held areas as the perimeter defences held by the British troops were pressed back and compressed ever-tighter into the town of Arnhem. The German mortar barrages could last for anything up to two hours and the British troops knew the Germans were experienced troops who knew precisely what they were doing. Any movement in the open attracted enemy fire, especially from mortar positions. Each day's action would begin at dawn with a mortar barrage, which the British troops referred to in typical sarcastic military humour as the 'morning hate'. This tactic had been used four months earlier by the 2nd Battalion, King's Shropshire Light Infantry (KSLI) during the Normandy campaign when several times a day they fired all available guns and mortars against the town of Lebisey, where the Germans held defences. The KSLI believed this 'one minute's worth of hate' was worth the effort. Indeed it was confirmed by surrendering troops who stated the tactic had the effect of disrupting normal daily routine. Albert Blockwell, who served with the King's Own Scottish Borderers at Arnhem, remembered this 'morning hate' all too well with 'mortar bombs dropping, 88mm shells ... and the never-ending *snap* of the snipers'. Despite the mounting casualties inflicted by these bombardments and other fire, they always mustered enough men to respond to an attack. One survivor recalled how on Sunday 24 September after 'a noisy night, with lots of shooting', the Germans opened fire with heavier weapons including mortars and rifle fire, during the course of which he was wounded and received 'some bits of mortar bomb in my shoulder'. Despite his wounds, he was still able to fight.

Finally, it was decided that the situation was untenable and the order was given to evacuate the troops from Arnhem. On the night of 24/25 September, using rubber boats, the survivors were taken back across the Rhine. Around 11,000 men had flown or parachuted into Arnhem but only 2,300 were evacuated. Although the British had managed to evacuate a number of men during Operation Berlin, they still found themselves being fired on by German mortars, as Sergeant Peter Quinn of the Reconnaissance Regiment remembered: 'A clatter of mortar bombs came down, lighting the woods and road with a queer blue light. The men scattered like demons in a pantomime. I was lifted off my feet with the blast of one bomb'.

Later Operations

At the time of the Allied landings in Normandy, the Soviets were biding their time and building up their resources in readiness for a series of massive offensives. They had been engaging the Germans in localised attacks to prevent them from guessing that anything was in the offing. Announcement of the Allies landings in Normandy was delayed on Soviet radio or in newspapers until 14 June, because Stalin wanted to be sure the Allied attack was a success. Finally, on 22 June, three years to the day after Germany had invaded the Soviet Union, the Red Army launched an offensive on an unprecedented scale. This was Operation Bagration, which had been in planning since April 1944 and was the culmination of all the other operations which the Soviets had been conducting since the end of 1943. For the Russians, 1944 would become known as 'The Year of Ten Victories'. The prelude to the first of these came on 14 January 1944, when a massive artillery barrage lasting just over an hour fired 100,000 shells on German forces investing the city of Leningrad. The following day an even-more intensive barrage lasting 105 minutes poured out 200,000 shells. The Germans were shattered and had nothing to respond with. The full weight of the Russian attack forced the Germans to pull back and after only twelve days of fighting the city was finally relieved, following a siege lasting 890 days. Armaments factories across the Soviet Union in 1944 produced 29,000 tanks of all types, 122,500 pieces of artillery and mortars, 40,300 aircraft of all types and 184 million shells, bombs and mines. Such an output allowed the Red Army to advance and push the Germans out of Poland and other occupied territories. As the Germans retreated in the wake of the Allied advances, they lost the armaments productions centres in the once-occupied countries of France, Belgium and Holland in the west, whilst in the east the factories in Poland were lost. This placed greater pressure on German armaments factories to produce more weapons and ammunition. These factories were being bombed by the USAAF during the day and the RAF at night, which caused severe disruption. Germany was also losing its allies one by one. On 24 August 1944, Romania surrendered, followed by Bulgaria on 9 September and finally Hungary on 15 October. The loss of these troops was compounded by the fact that armament factories in those countries were also gone.

Figure 32. German troops in action with a GrW34 8mm mortar on the Eastern Front.

On 17 February, at Korsun, where a force of 60,000 Germans had been holding out, the last of the resistance ended. The Red Army continued to advance at all points along the front and forced the Germans back to Romania, Poland and Norway. In preparation for Bagration, which concentrated on the region of Minsk, an area held by the German Army Group Centre, the Russians assembled an artillery force of 285 guns per mile. When they opened fire at first light, the T-34 tanks supported by SU-22 assault guns rumbled forward. The Germans held a position which bulged forward into a salient, allowing the Russians to attack from three sides; north, south and centre. The Red Army committed 2.3 million troops to the offensive along with 80,000 Polish troops, supported by 2,715 tanks, 1,355 assault guns and 32,968 pieces of artillery and mortars of 120mm calibre. Within three days the Germans had lost 20,000 killed and 10,000 captured, along with all their equipment and vehicles. The main fighting of Operation Bagration itself was concluded on 3 July, but sporadic combat continued until August. The battle had cost the Red Army over 178,000 killed, but they had inflicted 350,000 casualties on the Germans and captured a further 150,000 prisoners. The German Army in the east was now incapable of launching any credible offensive or counter-attack because of the weight of its losses in troops, weapons and equipment. The Red Army continued to advance and by January 1945 it was on the

Figure 33. German GrW36 5cm mortar in action on the Eastern Front.

eastern border of Poland. By March, the Red Army was poised on the German frontier ready to invade.

The Allies in the west had continued to advance and by December 1944 they were deep into Belgium. All the time the Germans showed that they were far from being a defeated force. They could regroup and mount counterattacks, which on occasion were surprisingly effective. Ammunition might be in short supply for the German artillery, but with careful planning sufficient shells could be conserved to allow heavy barrages to be fired. Artillery and mortar barrages were referred to by the British soldier in typical army slang as 'stonk'. In October 1944, the Suffolk Yeomanry, serving with the Gloucestershire Regiment in the area of Aerle in Belgium, came under one such bombardment. During the shelling, all the troops could do was take shelter until it stopped. Then the dead and wounded were gathered up for burial or taken to Casualty Clearing Stations. After one 'stonk' a veteran remembers 'the stretcher bearers collected their grim burdens and trudged back along the road with them to the waiting ambulances, while men lay gazing in wonder at an arm or leg that wasn't there any longer, and while their mates lay, blackfaced and still warm in death'. In return, the Germans were usually given a good 'stonking' by British mortars, such as the heavy night-time barrage fired against German positions at Geilenkirchen as the Allies approached the West Wall, or Siegfried Line. In mid-November 1944, Lieutenant General Brian Horrocks, commanding XXX Corps, ordered the town of Bauchem to be engaged. Eric Codling was serving with the 8th Middlesex Regiment and remembered the event as a 'nightmare chore carrying 80lbs of bombs 2/3 yards from the road, thousands of rounds, then transfer mortars into prepared pits'. He remembered that, between them, three platoons of mortars fired 10,000 rounds in a space of three hours and fifteen minutes at Bauchem. They were supported by Divisional artillery and, not unsurprisingly, when they advanced there was very little resistance encountered. This 'stonk' equated to 100,000lbs of 3in mortar ammunition or 44.64 tons of ammunition fired at the average rate of 51.28 bombs every minute for the 195 minutes of the barrage. Little wonder that resistance was light after such a pounding.

On 16 December, the Germans launched a surprise attack of unexpectedly strong proportions and caught the Allies totally unprepared. This was *Wacht am Rhein* (Watch on the Rhine), which Hitler had been planning for months. The preparations were well laid and included thirteen divisions with 200,000 troops, 340 tanks and 1,600 guns. With bad weather keeping the Allied air forces grounded, the Germans manoeuvred without fear of being attacked from above. American forces received the brunt of the attack and held out in defensive pockets of resistance such as the town of Bastogne, which was strategically vital to the success of the German operation. The town held firm despite all attempts to capture it, and eventually the weather cleared and Allied aircraft could deliver supplies to the beleaguered troops and attack German positions. By 25 January, the fighting, known to the Allies as the Battle of the Bulge, was at an end. German forces were severely depleted, having committed the best of their experienced troops to the offensive along with the last of their armoured reserves of any consequence. In March 1945, the Allies were in northern Germany and were able to launch Operation Varsity, the Anglo-American airborne crossing of the Rhine River, with some 40,000 troops being either parachuted or flown in by glider to seize positions. British gliders alone flew in almost 5,000 men, 342 Jeeps, hundreds of trailers and motorbikes, anti-tank guns and ten 4.2in mortars to serve as fire support along with the 75mm guns and 25-pdr field guns. On 25 April, the Allies linked up with their Red Army allies at Torgau on the River Elbe. Austria had been overrun and most of Czechoslovakia had been cleared of German troops. It was almost the end of the war in Europe, and after eleven months of fighting from June 1944, the US artillery had fired 1,496,000 tons of ammunition of all calibre at an average rate of 8 million rounds per month. The stocks of small arms ammunition and mortars also ran into the millions of rounds.

The last battle remaining was the fight to capture the German capital city of Berlin, which had been turned into a stronghold with rings of defences resembling those which had encircled the Soviet cities of Leningrad and Stalingrad. The logistical scale of the build-up for the attack on Berlin was immense. In 1944, Russian factories produced 29,000 tanks of all types, 122,500 pieces of artillery and mortars, 40,300 aircraft of all types and 184 million shells, bombs and mines. Zhukov noted the 'nature of the operation required a steady stream of ammunition from front depots to the troops, bypassing the intermediate links such as army and divisional depots.' Engineers converted the German railway line to the Russian gauge and ammunition was brought up almost to the banks of the River Oder. A fleet of 2,450 train wagons moved 1.23 million shells, totalling some 98,000 tons, forward in readiness. It was an enormous undertaking and it has been calculated that if the trains used to carry these supplies were stretched out they would have extended 746 miles. An army of engineers built bridges, repaired roads and railway tracks and cleared obstacles. The First Belorussian Army had 125 combat engineering battalions attached to it and these built 136 bridges for the advance and constructed command posts and bunkers in the thousands. The attack on Berlin would involve 85,000 trucks and 10,000 other vehicles to tow the artillery and heavy mortars. A total of 7 million shells were stockpiled in readiness and it was believed that over 1 million would be fired in the opening phase of the

Figure 34. Soviet 82mm mortar in action, probably during the battle for Berlin, 1945.

attack. By 15 April, units of the Red Army were only 40 miles from the centre of Berlin. In the centre, the 1st Belorussian Front commanded by General Zhukov took the city of Berlin as its axis of advance and pressed ahead 6 miles across the Oder River along a front 30 miles wide. Joining in the advance was 2nd Belorussian Front in the north and 1st Ukrainian Front in the south. The Americans remained at Torgau whilst the fighting for the city raged street by street. As they entered the city proper, the Soviet advance slowed down somewhat as the Germans put up a concerted defence, but it was only prolonging the inevitable. The Russians scheduled the final push for 16 April and the 2nd Belorussian Front was ordered north to deal with the German 3rd Panzer Army. Even so, this still left Zhukov with 2.5 million men, 41,000 pieces of artillery and mortars, 6,250 tanks and AFVs, with air support from 7,500 aircraft. Hitler committed suicide on 30 April, but units still continued to fight. On 8 May, the fighting finally ended completely as Germany surrendered unconditionally, just as the Allies had planned at the Arcardia Conference held in Washington between 22 December 1941 and 7 January 1942.

The Pacific and Far East

To defend the islands they had captured, the Japanese created a series of units called *Nanpo Shitai*, meaning Southern Area Detached Units. These were autonomous groups comprising of two infantry battalions supported by a small detachment of tanks, along with mortar platoons. In fact, all organisations within the Japanese Army had some form of mortar unit armed with a variety of calibres of weapons. For example, an Independent Brigade may have two companies of eight to sixteen mortars or a single company armed with either four or eight mortars of 81mm or 90mm calibre. The British Army and troops from the British Empire, including India, Australia and New Zealand, along with Americans, came to know the Japanese as tenacious fighters who would defend even the most desolate places and stubbornly fight to the very last. For example, when the 1st US Marine Division landed on the tiny island of Peleliu, measuring just 6 square miles, during Operation Stalemate, they fought a hard battle against a garrison of 11,000 men. When the last Japanese surrendered in February 1945, the fighting had cost them over 10,600 killed and barely 200 captured. To complete the taking of this island, American troops had fired more than 15 million rounds of small arms ammunition, 150,000 mortar bombs, and thrown over 118,000 hand grenades. It was later calculated that factoring in naval fire support it had taken an average of 1,500 rounds of artillery to kill each Japanese soldier and capture one small island. After the war, the US Ordnance Department recorded that 750,000 pieces of

Figure 35. The US Army 60mm M2 mortar was used in the Pacific and European theatres.

artillery of all calibres had been used by the US Army and Marine Corps during the war in the Pacific. These weapons had fired 11 million tons of ammunition and the mortars had fired a further 476,000 tons of ammunition.

The Pacific War was dominated by naval battles, with airpower being a major contributing factor in most cases. Several large-scale engagements decided the outcome of the war in the Pacific, and these would have a direct bearing on the way in which Japan conducted the war. As an island nation, Japan had to maintain a large maritime fleet for merchant trade in peacetime, and during the war this fleet also supplied the island garrisons spread out across thousands of miles of ocean, such as the Philippine Sea. The Imperial Japanese Navy expanded, with aircraft carriers, submarines and even huge battleships. Japan had no large reserves of natural resources and the reliance on shipping to bring in raw materials to the main island with its armaments manufacturing base was crucial. This included oil, coal, rubber and iron, all for the war effort. The Americans committed much of the Army and its equipment to North Africa and then the build-up for the inevitable invasion of Europe. This left the US Navy and Marine Corps to deal with the war in the Pacific, while British and Commonwealth troops, along with American support, fought the land war in the steaming jungles of Burma. Japan had gone to war with a merchant fleet of over 5 million tons, but the efforts of the US Navy reduced this to 670,000 tons by 1945. This meant the islands could not be supplied with even the most basic necessities. Japan could not hope to compete with America in terms of armaments production. Between 1943 and 1944, Japan launched seven aircraft carriers and in the same period America launched ninety aircraft carriers. By January 1944, the number of aircraft available to the Japanese air force for the theatre of operations was 4,000 machines, while the Americans had almost 11,500 aircraft of all types. The Imperial Japanese Navy lost 2 million tons of warships, which meant it could not provide escort protection to convoys sailing to the garrisons.

The US Navy fought its first major naval battle in the Coral Sea between 4 and 8 May 1942 and won a costly victory. Exactly one month later it fought another battle at Midway between 4 and 7 June, and won such a decisive victory that it broke the Japanese naval dominance in the Pacific. American planners decided to invade the Solomon Islands. In order to do so they would first have to capture the island of Guadalcanal, which lay to the east of the Solomon Islands. This was the first operation in a plan known as 'Island Hopping' and would involve a series of amphibious assaults on each island being attacked. Defending the islands in the group made up of Tulagi, Florida Island and Guadalcanal, the Japanese had deployed almost 4,000 men and 100 aircraft. The Americans assembled seventy-five warships and transports to carry the invasion force, which included 16,000 US Marines. At the time, American troops had no combat experience of landing in an amphibious force of such size. This was only six months after Pearl Harbor and the whole enterprise had been assembled very hastily. For example, the men had only sufficient ammunition for ten days and their supply levels had been reduced from three months to only two months. The first wave of Marines landed at Lunga Point on Guadalcanal on 7 August. The landing was largely un-opposed due to the bombardment by naval gunfire and air strikes.

The main reason to attack Guadalcanal was to prevent the Japanese using an airfield on the island to strike American shipping. As the Marines moved inland they overran the airfield on 8 August, and apart from some localised resistance no major counter-attack was made by the Japanese. The situation could not last and on 21 August a large Japanese force attacked and was destroyed. The Japanese were able to land reinforcements on the island and in early September they made a particularly heavy night attack on positions held by the US Marines. Using their mortars, the Japanese tried to isolate and divide the American positions. The Marines replied with their own mortars, which at one point were almost firing on their own troops. To try and dis-cover what the Americans were planning as their next move, the Japanese listened in on radio communication. The important orders were in code but some messages were transmitted in plain language, which English-speaking Japanese could understand. The US began to use Native North Americans who spoke in their tribal languages – incomprehensible to the Japanese. These natives included Navajo. They were known as 'Code Talkers' and were follow-ing in the tradition of other natives who had been used in this way during the First World War. The natives used codewords to describe what they saw, such as 'turtle' to mean tank. The Navajo language was very complex and their word for mortar was *Be-al-doh-cid-da-hi*, which meant 'sitting gun'. When they spotted the weapon, this was passed to another Navajo-speaking native who translated. This method was also used on Saipan, while on Iwo Jima, Seminole natives were used as Code Talkers. Other natives included Comanche and, in North Africa, Meskwaki Code Talkers baffled the Germans and Italians.

The fighting in the operation lasted until 7 February 1943, costing the Americans 1,752 killed and 4,359 wounded. During the operation, equipment and weapons was lost and supplies of ammunition also became critical at one point. The Marines quickly learned to improvise with anything to hand. During one particular firefight, a group of Marines became pinned down by Japanese fire. Joining their group was a Marine from one of the mortar sec-tions armed with an M2 60mm weapon but only four rounds of ammunition. He had lost contact with other members of his section and the bipod and baseplate had also been lost. Undeterred, he brought his weapon into action by jamming the base of the barrel into the ground, supporting the barrel with one hand and loading with his other hand. The four rounds he fired helped subdue the Japanese position just long enough for the Marines' artillery to fire in support of the action. Holding the barrel by hand in order to fire the weapon was not a problem, because the British 2in mortars and Japanese Type 98 'knee mortar' were fired in such a way. The logistic support for the Guadalcanal operation had been confused but eventually it was resolved and much was learned from the experience. The landings were eventually supported by M3 Stuart light tanks, which the Japanese had little to directly oppose. The tanks literally crushed everything beneath them as they attacked the Japanese positions. Major General Alexander Vandegrift, commanding the 1st US Marines on Guadalcanal, recorded how the 'rear of the tanks looked like meatgrinders' from the bodies they had driven over with their steel tracks. It was the first operation in the Island Hopping campaign and every shell, vehicle, gallon of petrol and soldier had to be transported in readiness for each such amphibious landing.

Like all other troops, the US Marines knew the usefulness of mortars in providing fire support, especially during amphibious landings. A rifle company of US Marines included an eleven-man mortar section comprising two squads, each armed with an M2 60mm calibre mortar. The mortar platoon had four M1 81mm calibre mortars. Battalion commanders came to regard these as their own 'artillery' which could be used to fire HE bombs to break up Japanese attacks, known as *Banzai* charges, and to destroy defensive positions. The British and American troops learned to concentrate their fire against these fearsome charges and the Japanese, in turn, learned to bring their own mortar and machine-gun fire to bear in support of them. In addition to the 60mm and 81mm calibre weapons, the Marines also had the heavier 4.2in calibre mortar which was used for the first time in August 1942, when the 1st US Marine Division landed on Guadalcanal, and was deployed again two months later when landings were made on Bougainville. Mortars proved invaluable in the jungle terrain on these islands, remaining operational when heavier weapons could not cope or were unavailable. The United States Marine Corps (USMC) reached a peak strength of 450,000 men in 1945, a massive increase from its pre-war level of fewer than 20,000 men. Total casualties would amount to 91,718 killed, wounded and taken prisoner in some of the fiercest fighting of the war. The USMC also raised an airborne unit known as the 1st Parachute Regiment, which was formed in 1941 and by early 1943 had four battalions. Despite its status as a parachute regiment, the troops did not deploy from aircraft but were used as elite infantry. The regiment's first engagement was in 1942 when it confronted Japanese troops in the Solomon Islands. The 2nd Battalion was deployed to the island of Choiseul on 28 October where, along with Australian troops, they were used to raid into Japanese positions over a period of five days. The garrison was estimated at between 3,000 and 7,000 soldiers. They were defending an island of some 1,147 square miles. Although the attacking force was only 750 men, over a period of two days it managed to destroy stores before moving its attention to the tiny Guppy Island, which was the main supply base for the larger island. The attack included bombarding the base and supply dumps with mortar fire, which was successful in destroying stocks of ammunition and food. The Japanese were convinced this was a preliminary attack in preparation for the main assault. However, it was actually a diversionary attack and the main assault went in at Bougainville. By 3 November, the Americans and Australians had left Choiseul but they had shown how even a lightly-equipped unit was able to use its mortars to inflict damage beyond its size in a way which would ordinarily have been designated to tanks or artillery.

Because the Marines were amphibious landing troops, it was proposed they should be supported on operations by boats or landing craft specially converted and fitted with mortars, which would serve as floating battery

Figure 36. Japanese 81mm mortar in action. Note the flag on the ground used to identify the position to Japanese aircraft.

platforms to provide covering fire. The idea had been proposed for the landings on Sicily in July 1943 but the craft were not ready in time for the operation. The US Navy applied itself to helping out its Marine cousins and by 1944 had produced a number of floating mortar platforms by fitting three 4.2in mortars on Landing Craft Tank (LCT) which were used during the Island Hopping campaign in the Pacific. Each vessel had three weapons mounted to fire forward and space had been converted to provide magazine storage for 3,200 bombs and sufficient room for the crews of the weapons. They were prepared to be deployed to cover the landings on Saipan by the 2nd and 4th Marine Divisions in June 1944.They were loaded onto transport ships to transport them to their area of operations, but two were lost when they were washed overboard during a storm. A third was lost when its transport ship blew up. The remainder of the mortar-carrying LCTs were eventually deployed to support the 1st Marine Division during the invasion of the Palau Islands in September 1944. The floating batteries provided fire support and could move along the coast to fire on new targets. These vessels continued to be used operationally in this role until the end of the war.

Each amphibious landing took the fighting closer to the Japanese home islands, and the fighting became stronger as the Japanese tried to resist the American advance. On Tarawa the island was held be a combined force of 5,800 troops and labourers. The Japanese commander, Keiji Shebaski, had boasted that it would take a million Americans 100 years to capture the island. In the end it took 35,000 US Marines just three days between 20 and 23 August 1943. They lost 1,009 men killed and 2,101 wounded, whilst the Japanese garrison lost 4,690 killed with only 146 taken prisoner. Saipan was assaulted on 15 June 1944 and was defended by some 3,000 Japanese troops. The US Marines committed 71,000 troops to the attack. The fighting lasted until 9 July, by which time they had sustained 3,426 killed and 10,364 wounded. Only 921 Japanese were taken prisoner. Next was the invasion of Tinian on 24 July 1944, when a force of 41,364 US Marines went ashore. The fighting lasted until 1 August, by which time the Marines had lost 326 killed and 1,593 wounded. Of the Japanese force only 252 surrendered, with 5,542 killed and the remainder being listed as missing. Iwo Jima was the first major assault of 1945, with the first of a force of 70,000 US Marines landing on 19 February. The Japanese garrison of 22,000 troops had built extensive defence works which had been laid out to provide overlapping arcs of support fire with machine guns and mortars. By the time the island was captured on 26 March, the US Marines had lost 6,821 killed and 19,217 wounded. Some 18,844 Japanese were killed, many were missing and only 216 surrendered.

In Burma, over the same period, the British and Commonwealth troops from India and Australia were engaged in fierce jungle fighting and pushing the Japanese back. Mortars proved to be so valuable in the Burma campaign that some anti-tank gun units were converted to use 3in mortars to provide fire support. This was possible because Japanese tanks were very light and vulnerable to light anti-tank weapons. Also, tanks were not suited to jungle warfare and not many were encountered.

The island of Okinawa lies only 340 miles from the main islands of Japan, meaning that its capture would allow aircraft to fly unhindered and bomb the main island cities and factories. The US attack was launched on 1 April 1945 with a force which would see some 183,000 troops committed. The Japanese defence force comprised 80,000 regular troops and 40,000 Okinawan conscripts. Fighting was tough, the expenditure of ammunition high and weapons needed replacing. On 9 May, a request was sent out to Oahu calling for a resupply of 100 mortars of 81mm calibre and thirty mortars of 60mm calibre. At the same time the entire reserves stock of twenty-seven 81mm mortars was requested to be sent from Saipan. These were sent immediately and arrived on Okinawa four days later. Usage of mortars was particularly high and by mid-May XXIV Corps had fired almost 269,000 rounds of 81mm ammunition. On 21 May, a request for 100,000 mortar bombs was put in but stocks were low. Air Transport Command ferried what stock were immediately available, which amounted to 36,000 rounds. By the time the fighting ended on 22 June, the US assault force had lost 38,000 killed and wounded. The Japanese had lost 110,000 killed and 7,000 taken prisoner. On 6 August, the first atomic bomb in history was dropped on Hiroshima and exploded with a force previously unimaginable. This was followed up on 9 August by the second atomic bomb – dropped this time on Nagasaki – which exploded with a force equivalent to 22,000 tons of TNT. Finally, on 2 September 1945, Japan surrendered.

Chapter Nine

Self-Propelled Mortar Carriers

Half-track carriers were one of the most versatile designs of all armoured fighting vehicles to be used during the Second World War. The Japanese Army had this type of vehicle, as did the French Army, but it was the German and American armies which developed their half-track vehicles to serve in a whole range of roles, from mounting anti-tank guns and field guns to serving as carriers for mortars. One of the first types to be developed for the mechanised infantry battalions of the US Army was the M4, which entered service in October 1941. It carried an M1 81mm mortar in a fixed mounting to allow it to fire rearwards from the back of an M2 half-track vehicle. Unfortunately this layout was not favoured, probably because the carrying vehicle had to be manoeuvred into firing position instead of simply being driven forward to open fire on targets, like standard self-propelled guns such as the M7 'Priest' with its 105mm gun. A modification was made so that the crew could dismount the mortar in order to fire it on a baseplate from prepared weapon pits. The modified mounting corrected the drawback and fitted the mortar to allow it to fire forward from within the vehicle. It was operated by a crew of six men and carried ninety-six rounds for the M1 mortar, which comprised mainly HE but with some smoke and illuminating bombs. Between late 1941 and December 1942, the White Motor Company of Cleveland, Ohio, produced 572 of these vehicles, which went on to serve in mainly the European theatre. The design weighed 7.75 tons, had an overall length of 19.72ft and could reach speeds up to 45mph on roads. It measured 6.43ft in width and 7.4ft in height and carried a .30in calibre machine gun for self-defence with 2,000 rounds of ammunition. Some vehicles were armed with the heavier .50in calibre machine gun, and the crew also had personal weapons.

Another variant was designated as the M4A1, and from May 1943 the White Motor Company built 600 of these vehicles. This was slightly larger and heavier weighing 8 tons but still carrying ninety-six rounds of ammunition for the M1 81mm mortar, which was mounted to fire forward. A crew of six operated the vehicle and weapons, which included a .30in calibre machine gun with 2,000 rounds mounted for self-defence. The M4A1 was 20.3ft overall in length, 7.44ft in height and 6.43ft in width. It could reach speeds of up to 45mph on roads. Together with its M4 counterpart, these mortar carrying

Figure 37. US Army M4A1 carrying an 81mm mortar. Note the ammunition storage on the side. The same amount was stored on the other side.

vehicles served with armoured units such as the 2nd Armoured Division, nicknamed 'Hell on Wheels', from 1942 and later served across Europe after June 1944. Despite the successful development of these two types of mortar carrier, the Ordnance Department decided to re-evaluate the layout and develop a third type of mortar-carrying half-track based on a modified M3 half-track and conduct experiments with an 81mm mortar mounted to fire forward over the driver's cab.

Field trials and firing tests proved this new layout to be superior to the M4 design in some respects, and in June 1943 it was standardised as the M21. The White Motor Company, with its experience in developing such vehicles, was awarded the contract to build the new design, and between January and March 1944 produced 110 units. Meanwhile, trials were continuing using an M4 half-track to mount a 4.2in (107mm) mortar for use with the chemical mortar battalions. Mobility and firing trials were conducted to assess the feasibility of this combination to lay smoke screens. The mounting was the same as that used on the 81mm mortar but the recoil forces of this heavier weapon proved too great for the vehicle's chassis, the trials were suspended and the project dropped. Two other projects, known as T27 and the T27E1, using the M1 mortar mounted in the chassis of tanks, were examined, but these were terminated in April 1944. The T29 to mount an 81mm mortar into a converted chassis of an M5A3 light tank was another short-lived project which never got off the drawing board. The Ordnance Department then tried mounting the 4.2in mortar on the M3A1 half-track, and this proved much better. For some reason the design team appears to have reverted to mounting the mortar to fire rearward out of the vehicle and the configuration was designated T21. A change of design to mount the mortar to fire forward resulted in the designation T21E1, and even mounting the weapon into a the chassis of an M24 light tank was considered, but it was not pursued and the complete project was dropped shortly before the end of the war in Europe in 1945. Two other proposals for self-propelled mortar carriers were the T36 and T96 projects. The T36 suggested mounting a 155mm mortar in the chassis of an M4 Sherman tank and the T96 a 155mm mortar onto the chassis of the M37 gun carriage. They were good ideas but by the time these proposals were put forward the war was coming to an end and the projects were dropped.

The M4, M4A1 and M21 mortar carriers were based on the M2, M2A1 and M3 half-tracks respectively, of which some 60,000 of all types were built. They served in various roles, including self-propelled gun and anti-aircraft gun platform with quadruple-mounted .50in calibre heavy machine guns known as the M16. There were also communications vehicles in this range. The White Motor Company built the prototype of the M21 in early 1943 as the T-19 and, following successful trials, it was standardised in July the same year. It was accepted into service in January 1944 and among the units to receive the vehicles was the 54th Armoured Infantry Regiment of the 10th Armoured Division, which later saw heavy fighting during the Battle of the Bulge in December 1944. The M21 had a crew of six to operate the vehicle, mortar and the machine gun for self-defence, while frames on the side of the vehicle allowed mines to be carried which could be laid for defensive purposes in an emergency. The vehicle had a combat weight of 20,000lbs (almost 9 tons) with an overall length of almost 19ft 6in. The height was 7ft 5in and it was almost 7ft 5in at its widest point. The barrel of the M1 81mm mortar was supported with a bipod and a special baseplate mounting which allowed it to be fired from the rear of the vehicle. A total of ninety-seven rounds of ammunition were carried and included smoke, illuminating and high explosive rounds. A store of forty rounds of ammunition was kept in lockers either inside the hull where the crew could access it easily ready to use. A further fifty-six rounds were kept in storage lockers, twenty-eight rounds either side of the hull, which could be loaded into the rear of the vehicle to maintain levels of ammunition ready to fire. This arrangement was the same on the M4 and M4A1 vehicles.

Figure 38. US Army 81mm mortar mounted on M4A1 half-track. A total of twenty rounds of ammunition were carried on either side of the weapon in the rear of the vehicle.

The mortar of the M21 could be traversed 30 degrees left and right; for greater changes the vehicle had to be manoeuvred to face the direction of the target. The mortar could be fired at the rate of eighteen rounds per minute to engage targets at ranges of almost 3,300 yards with the high explosive rounds. The barrel could be elevated between 40 and 85 degrees to alter the range. The .50in calibre machine gun was fitted on a pedestal mount to the rear of the vehicle and a total of 400 rounds of ammunition were carried. From there the firer could traverse through 360 degrees to provide all-round fire support. The vehicle was only lightly armoured up to a maximum 13mm thickness.

The M21 was fitted with a White 160AX six-cylinder petrol engine which developed 147hp at 3,000rpm to give speeds of up to 45mph on roads. Fuel capacity was 60 gallons and this allowed an operational range of 200 miles on roads. The front wheels were operated by a standard steering wheel and the tracks were fitted with double sets of twin bogies as road wheels, larger 'idler-type' wheels at the front and rear of the track layout and only one return roller. The open top of the vehicle could be covered by a canvas tarpaulin during inclement weather and this could be thrown off quickly when going into action. Although only few in number, together with the more numerous M4 and M4A1 mortar carriers, the three designs provided excellent mobile fire support to infantry units wherever required. All three designs were equipped with radio sets to communicate and receive orders as to where to deploy if needed to fire against targets. Some units of the Free French Army were supplied with some fifty-two examples of the M21 self-propelled mortar vehicles, which were used during the European campaign.

One armoured unit, the 778th Tank Battalion, recorded of the mortar carriers attached to D Company in December 1944 that the fire support they provided was 'instrumental on several occasions in assisting the advance of the infantry by placing fire on enemy gun positions and strongpoints that could not effectively be fire upon by other weapons'. The account continues by stating how 'the two ... mortar platoons, from advantageous positions on the west side of the Saar River placed harassing fire on the city of Bous, on the east side of the river. The platoon fired an average of 350 to 400 rounds per day into the city'. Continuing in their support of D Company, the mortar carriers fired from elevated positions at Bisten from where they suppressed German positions. Another armoured unit, the 746th Tank Battalion, was provided with fire support from mortar carriers and the unit recorded how these vehicles were able to 'fire support to [cover] advance infantry elements in many instances when tank fire cannot be employed successfully'. This account continues by recording how self-propelled mortar carriers 'were attached to an infantry regiment and further attached to one battalion and the assault company thereof. By following closely behind the advancing infantry, the mobile mortars lay down covering fires within their maximum range before displacing to the next bound. In some actions, the mortar carriers have backed down the axis of advance from one bound to another.' Yet despite the mortar carrier's effectiveness in supporting advances at very close quarters and keeping up with the advance, by the end of the war some officers in armoured units dismissed their usefulness. There were plans to develop the M21 vehicle to carry the larger 4.2in calibre mortar but it never entered service.

During its rearmament programme the German Army investigated the possibility of using half-tracked vehicles, and the way in which they could be developed into a variety of roles to support troops on the battlefield. By the time Poland was attacked, the German Army was equipped with several versatile designs of armoured half-tracked vehicles, mostly serving in the primary role of transporting troops on the battlefield and a secondary role as communications vehicles. Production of these designs continued so that several months later, when the *blitzkrieg* was launched against Western Europe in May 1940, the fleet of half-track vehicles was even larger. The two most widely-used types were the SdKfz 251 and the smaller SdKfz 250, which went on to prove itself to be no less versatile than its larger counterpart. In fact, by the end of the war in 1945; the SdKfz 250 had been developed into no fewer than twelve different configurations.

The German Army was quick to realise that light armoured half-track vehicles could be used on the battlefield as flexible workhorses. Of all the designs to enter service, it was the SdKfz 251 series, weighing 8.7 tons in its basic APC version and capable of carrying ten fully-equipped infantrymen as well as the driver and co-driver, which would prove invaluable in many campaigns, including North Africa. From the very beginning it complied with the requirements calling for an armoured vehicle capable of transporting infantrymen on the battlefield. Known as the *Gepanzerter Mannschraftstran-portwagen* (armoured personnel carrier) when it was first proposed in 1935, the vehicle quickly took shape and in 1938 the prototype was ready for field trials. It was produced by the companies of Hanomag and Bussing-Nag, which built the chassis and hulls respectively, and the vehicle was given the title of *Mittlerer Schutzenpanzerwagen* (medium infantry armoured vehicle) with the designation of SdKfz 251. The first vehicles were in service in 1939 and some were used during the campaign against Poland. Production was low at first, in fact only 348 were built in 1940, but there were enough numbers to be used during the campaign in the west in 1940. The SdKfz 251 was fitted with a Mayback HL42 TKRM six-cylinder water-cooled petrol engine which developed 100hp at 2,800rpm to give road speeds of up to 34mph, which was more than sufficient to keep up with the tanks in the armoured divisions.

The APC version was 19ft in length, 6ft 10in in width and 5ft 9in in height. The vehicle could cope with vertical obstacles up to 12in in height, cross ditches 6ft 6in in width and had an operational range of 200 miles on roads. Armour protection was between 6mm and 14mm, but the rear crew compartment where the infantry sat had no overhead protection, which exposed the troops to the elements and also the effects of shells exploding overhead. Two machine guns, either MG34 or MG42, were fitted to allow one to fire forwards from behind a small armoured shield and the weapon at the rear was fitted to a swivel mount to provide fire support for the infantry as they exited the vehicle. Being open-topped, the infantry could jump over the sides to leave the vehicle or exit through the double rear doors. The machine guns, for which 2,000 rounds of ammunition was carried, could be taken from the vehicle when the infantry deployed.

The SdKfz 251 was developed into a range of different purposes, from ambulance duties to anti-tank roles. By late 1944, around 16,000 vehicles had been built to serve in no fewer than twenty-three different roles. Depending on the role, each version had a different length of service life, but if they were capable of continuing to operate they remained in use. In fact, examples were still in operation right until the last days of the war at a time when fuel was extremely scarce. One of the earliest variants to be produced was the SdkFz 251/2, which was the mortar-carrying version, weighing 8.64 tons and equipped to carry the 8cm GrW34 mortar. Being open-topped, the weapon could be fired from within the vehicle, firing forward, and a separate baseplate allowed it to be dismounted for use from prepared positions. The vehicle in this role was operated by a crew of eight, available in the heavy platoon and known as *Great* 892 (Equipment 892). It carried sixty-six rounds of ammunition ready to use and was supported in turn by the SdKfz 251/4 version, which could carry resupplies of ammunition or even tow the heavy GrW42 12cm mortar.

The other half-track vehicle developed into a mortar carrier was the SdKfz 250, which was built by the company of Bussing-NAG, which developed the armoured body, and several other manufacturers including Wegmann and Deutsche Werke. Although the design had been thoroughly tested in the field throughout 1939, there were insufficient numbers ready to enter full operational service on the outbreak of war. In fact, the SdKfz 250, originally referred to as *Leichte Gepanzerte Kraftwagen*, did not enter service with the German Army properly until 1940, by which time it was known as the

Leichte Schutzenpanzerwagen (light infantry armoured vehicle). Although it was not in service for the Polish campaign, there were sufficient numbers in service to be used during the attack against Holland, Belgium and France, where they were used in roles such as reconnaissance, command and communications. After this initial battle-proving deployment, the SdKfz 250 went on to see service on all fronts during the war, including North Africa, Italy and Russia.

The basic model was an armoured personnel carrier designated as the SdKfz 250/1, operated by a crew of two (driver and commander). In this role it was capable of carrying four fully-equipped troops with support weapons, such as crew for mortars or machine guns. This version was armed with two machine guns, such as the MG42, for which some 2,000 rounds of ammunition were carried. The basic version SdKfz 250 was almost 15ft in length, 6.4ft in width, but the height varied according to the role in which it was serving and the armament carried. The standard version had a combat weight of 5.5 tons, but, again, this varied according to armament and other equipment, such as the mortar carrier which weighed 5.61 tons. The armour thickness was from 8mm minimum to 15mm maximum.

The vehicle in all its variants was powered by a Maybach HL42 TR KM six-cylinder water-cooled inline petrol engine, which developed 100hp at 2,800rpm and gave a top speed of just over 40mph on roads. The vehicle had an operational range of over 180 miles on roads and it could negotiate vertical obstacles up to 15in, ford water obstacles shallower than 27in and scale gradients of 40 degrees. The front wheels were not 'driven', being used for steering purposes only. The automotive power was to the front drive sprockets on the tracks and the suspension was of the FAMO type, and whilst the vehicle was itself efficient it was somewhat complicated to maintain. This was a telling point in the sub-zero conditions on the Russian Front after 1941. In total, twelve variants were developed from the basic version and included an anti-tank gun version, specialist engineer versions, signals vehicles, ammunition carrier with ordnance troops and was even used by the *Luftwaffe*. Most, but not all, versions of the SdKfz 250 were open-topped, which was perfect to allow fire support weapons such as the GrW34 8cm mortar to be mounted and create a variant known as the SdKfz 201/7 or *Great* 897 (Equipment 897). These were operated by a crew of five and available to the fourth platoon of the *Leichter Panzer Aufklarungs*, or light armoured reconnaissance vehicles. A total of forty-two rounds of ammunition were carried on the vehicle ready to use and the mortar could be dismounted to be used to provide fire support in prepared positions. It was supported by a version termed *munitionsfahrzeug* (ammunition vehicle), which was operated by four men and carried sixty-six

Figure 39. Captured French AMR35 tank converted to carry a GrW34 8cm mortar firing forwards.

rounds of ammunition to resupply the mortar carrier. It was armed with two machine guns for self-defence with 2,000 rounds of ammunition.

In addition to using its own standard mortar-carrying vehicles, the German Army also converted a number of captured French armoured vehicles to the role of self-propelled mortar carriers. These they armed with either German service 8cm sGrW34 mortars or captured French-built Brandt weapons. French tanks such as the AMR35 had their turrets removed and the chassis converted to other roles such as self-propelled guns and mortar carriers. In May 1940, the French Army had about 200 AMR35 tanks in service, mostly armed with a 37mm gun in a fully traversing turret, but there were also other variants. After the French surrender, all surviving examples and variants captured by the Germans were converted into other uses, which included carrier vehicles for the standard German Army 8cm sGrW34 mortar. In this role they were designated as 8cm *schwere Granatewerfer 34 auf Panzerspahwagen* AMR35(f), which identified it as a heavy armoured self-propelled mortar carrier.

The conversion was achieved by first removing the upper superstructure including the engine covering, and this was replaced by an open-topped fighting compartment into which was mounted an 8cm GrW34 mortar fitted on a race-ring mounting to allow it to be fired in any direction without having to manoeuvre the vehicle. The rear of the compartment was open but could be closed off with a door to protect the crew. The conversion gave it a larger profile than the original design but the engine and all other automotive parts and road wheel layout remained unchanged. This gave the vehicle a road speed of 31mph and an operational range of 120 miles. It was operated by a crew of four, which included the driver, and a secondary armament of a single 7.92mm calibre MG34 machine gun was fitted for self-defence. A supply of ready-to-use ammunition was carried on the vehicle and resupply vehicles would have brought forward replenishment stocks. Records show that around 200 such vehicles were converted to this role and used only in France, where they could be deployed in response to threats. The conversion would have been completed at workshops in France, but it is not clear if any of these vehicles participated in the fighting after the Allied landings in Normandy from 6 June 1944 onwards.

It would seem likely these would almost certainly have been deployed at some point against the Allies because it would make no sense to develop such weapon systems and not use them. It may be that some of these self-propelled mortar vehicles were used in the fighting during the Normandy campaign, but due to the low production numbers they have been overlooked in favour of the more widely-used vehicles such as the true self-propelled guns and tanks. The person responsible for developing these systems and other self-propelled weapons was Major Alfred Becker, who was a professional soldier, having served in the First World War. He was an engineer who excelled in developing hybrid systems such as these, using captured stocks of enemy equipment. He commanded the *Sturmgeschutz-Abt* 200, equipped with self-propelled guns of his design, as part of the 21st Panzer Division, seeing much action in Normandy. He served with distinction and developed other systems until his capture in December 1944.

One of the most unusual conversions to serve as a mortar carrier was based on the French SOMUA MCL half-track personnel carrier, which became known as the *Mittlerer Schutsenpanzerwagen* S307(f), work on which began in 1943. The vehicle was modified to its new role by mounting two rows of eight barrels of 81mm captured French Army mortars stacked on a mounting to the rear of the vehicle. The tubes were pre-loaded and could be fired simultaneously to produce an instant bombardment. Reloading the tubes would have taken time and rate of fire would have been a lot slower than using a

Figure 40. Detail of GrW34 8cm mortar mounted on a captured French AMR35 tank.

conventional mortar firing from a prepared position. In total some sixteen of these vehicles were available in 1944, but their fate is not known. A heavier version was produced, also based on the SOMUA MCL, which mounted twenty barrels of 81mm Brandt mortars in a similar array, and this was known as the *Schwerer Reihenwerfer auf* SPW SOMUA S303(f). These vehicles served in France, but, again, it is not known conclusively if they were deployed in action against the Allies after June 1944.

Captured armoured vehicles were dispatched to various theatres of operations, including Finland and Norway. Others such as the French Char B-1 bis, known in service with the German Army as the *Panzerkampfwagen* B-2 740(f), were sent to the Channel Islands and the Eastern Front, which represented the opposite extreme edges of the territory under German occupation. In the Channel Islands, some French tanks had their turrets removed to be incorporated into defence plans. Other vehicles which could have been converted to use as mortar carriers included the UE630, which the French Army used as a transport vehicle for supplies, and the *Unic-Kègresse* half-track, yet despite their suitability neither these nor apparently any other French vehicles were armed to serve as self-propelled mortars carriers.

The French Army had never deemed it necessary to develop a self-propelled mortar system using any of the weapons in service; after all they had good artillery and armoured units. The Italian Army did not develop a self-propelled mortar system and relied on artillery and the mortars used by the infantry units. After 1943, when Italy capitulated to the Allies, the German Army seized many armoured vehicles and took these into service. Unlike the French vehicles which they converted to other uses, the Italian vehicles were used in their primary roles. The Soviet Red Army did not develop a self-propelled mortar system either and relied on self-propelled guns and vehicles which towed the heavy calibre mortars on wheeled carriages. The Japanese Army did experiment with self-propelled mortars for a while and developed the Type 4 Ha To. This used a Type 4 Chi-To medium tank which was converted to allow a 300mm calibre Type 3 heavy mortar to be mounted to fire forward. It could fire a 374lbs HE bomb out to ranges of 3,300 yards, but the design was unstable and the vehicle proved liable to toppling over due to its height. In the end only three prototypes of the Ha To were produced and these never saw combat service.

British and Commonwealth forces did not show much interest in developing a self-propelled mortar version based on an armoured vehicle design. Some feasibility experiments were conducted to examine the viability of producing such a variant, but ultimately the research did not lead to the introduction of a vehicle-mounted mortar in the same way as used by the American and German armies. One experiment which did lead to the production of a self-propelled version of the British 3in mortar was based on the Universal Bren Gun Carrier. This was developed by the Australian Army, which had already modified some Universal Carriers to mount 2-pdr anti-tank guns, and using this as a starting point they fitted a 3in mortar to the vehicle. The weapon could be fired directly from the vehicle or dismounted and used from a prepared defensive position. The mortar had a full 360-degree traverse capability if required, which meant the vehicle did not have to manoeuvre to alter the range of traverse beyond the angles which could be achieved in the bipod mounting. In terms of range, this configuration was comparable to the standard infantry mortar and fired the same bombs. In the end it was never taken into service with the Australian Army, but around 400 examples are understood to have been produced and these were sent as part of the military aid to support the Nationalist Chinese Forces of Chiang Kai Shek in the fighting against the Japanese.

It seems strange that the British Army should not pursue the development of a self-propelled mortar vehicle, especially when it developed a range of specialist vehicles for other roles to clear minefields and close support tanks armed with large calibre guns. These were developed in the build-up for the invasion of Europe to support the landings. They were known as 'Hobart's Funnies', after Major General Percy Hobart who thought up some of the designs. Hobart was a military engineer who had served in the First World War, seeing action in France. During the 1920s, he developed an interest in tank designs and armoured warfare tactics. He retired in 1940 under duress, following a conflict of opinions concerning his designs for armoured vehicles and their role. Hobart initially joined his local Home Guard unit, but in 1941 he was re-instated and given the job of training the 11th Armoured Division. Further positions followed and in 1942 he was given the role of training the newly-created 79th Armoured Division. After the disastrous failure of Operation Jubilee, the Allied attack against Dieppe on 19 August 1942, where none of the tanks were able to get off the beach, he set about developing a series of specialist armoured vehicles designed to support future amphibious landings. What Hobart developed included bridge-laying vehicles, flails to breech minefields and flame throwers. These were to prove vital during the D-Day landings and campaigns across Europe. For some reason self-propelled mortars were not developed. One can only conclude that with SPGs such as the Sexton, with its 25-pdr field gun, and the M7 'Priest', with its 105mm gun, Hobart did not feel it necessary to build a design around the British 3in mortar.

Chapter Ten

The Weapons

At the start of the war, the British Army had two types of mortar in service, the smaller of which was the 2in weapon for use at platoon level within the infantry battalion structure. The original design on which the 2in mortar was based had been developed in Spain and the modified version which emerged from this entered service in 1937. In the early 1930s, the British Army was looking for a light mortar to replace the ageing weapon designs left over from the First World War. Learning of a new 50mm calibre mortar being developed in Spain in 1934, the British Army decided to inspect this weapon, thinking it might be what they were seeking. The calibre of 50mm was similar to many types then in service with other European countries, including France, and as it turned out the Spanish weapon was actually based on a Brandt design. As interesting as the weapon was, it was not exactly what the British Army wanted and they rejected the design as it stood, and made recommendations for some modifications which would make the weapon acceptable for the British Army.

At this time the Birmingham Small Arms Company had developed a 2.5in calibre mortar and there were also a few other designs being developed. Each of these were rejected in turn following firing trials, and the Armament Research Department again looked at the Spanish mortar, this time deciding to use it as a starting point from which the Armament Research Department could begin developing its own version. What emerged was the 2in calibre mortar. In November 1937, ten examples of the new weapon were prepared for firing trials, along with 1,600 rounds each of high-explosive and smoke bombs. The resulting trials confirmed the reliability and dependability of the weapon. In the face of growing political tensions across Europe, including Germany's continued rearmament programme, in February 1938, only four months after the firing trials, the Director of Artillery ordered the weapon to be placed in production. By early 1939, some 500 units of the weapon along with stocks of ammunition were entering service as the Mk II, and crews were being trained in the use of the new weapon.

Spain had also developed other mortars which were used during the Civil War, including the 8cm Mod 1933, which was actually 81mm calibre, and

Figure 41. Stokes trench mortar in collection of SASC.

weighed over 137lbs in action, with its barrel capable of being elevated between 30 and 85 degrees to assist in varying the range. It fired two types of bombs weighing 8.8lbs and 18.9lbs out to ranges of 3,300 and 1,750 yards respectively. The 12cm 'Franco' was the heaviest weapon, weighing 1,525lbs in action, and could elevate its barrel between 45 and 70 degrees. It could deliver a high explosive bomb weighing more than 36lbs out to ranges of around 7,000 yards. Although the country would remain neutral throughout the war, it was an unknown quantity at the time and the Allies could never be entirely sure that General Franco, whose Nationalist forces had been victorious in the Spanish Civil War, would not join forces with Hitler and Mussolini in return for them having supported his cause in the war. The British could not afford to relax their guard when it came to monitoring Spanish movements, especially on Gibraltar. During the war, the Spanish Army maintained a strength of 750,000 troops, and along with other weaponry had three main types of locally-manufactured mortars such as the 5cm Ecia Valero, which was produced by the company of Esperanza y Cia of Vizcaya, weighed 15lbs in action and the barrel could be elevated between 45 and 80 degrees. It could fire a 1.5lbs bomb out to a range of 1,100 yards and although the bomb was not very heavy it provided some degree of fire support and was actually considered to be one of the finest types in its category of light mortar for use at platoon level.

During the course of the war, the 2in mortar would be developed into no less than eight separate marks, from which also stemmed a number of other variations. Some were successful and others less so, such as the Weston version – developed in 1944 – which was found to be less than satisfactory on soft ground and was difficult to use. This version had the advantage of being fitted with an automatic re-cocking feature to the firing mechanism, but was still withdrawn from use. The standard service version of the 2in mortar had a barrel length of 21in and could fire a typical high explosive bomb weighing 2.25lbs out to a range of 500 yards. Originally it had been fitted with a large collimating sight with elevating and cross-level bubbles, but this was later omitted because it was considered unnecessary for such a light weapon in service with a front line unit. It was replaced with a simple white line painted up the length of the barrel. The firer only had to line this up in the direction of the target and fire a number of bombs for effect. Whilst this method of operation may sound rather ambiguous, it worked well and experienced firers could calculate the angles required for the bombs to hit the target area. The mortar evolved in other directions too, with the original large base plate being replaced by a simple curved plate, to give it a combat weight of 10.25lbs, with the barrel being supported by the firer's free hand. A version was developed with an extended baseplate known as the 'Palestine Base' and an example exists in the SASC Collection at Warminster. This is dated 1937, but since the 2in mortar did not enter service until 1939, this would be incorrect. Furthermore, it is quite possible this modification was done locally since it is not a standard design.

The 2in mortar could be fired at the rate of some eight rounds per minute. The illuminating round weighed 1lb and the smoke round 2.25lbs. A whole range of other ammunition was also developed, including a specialised bomb that cast a lightweight explosive-filled net over patches in minefields so that it could be detonated to clear a path. Versions of the weapon itself included the Mk VII* with a shortened barrel for use with airborne units, the Mk VII for use in Universal (Bren Gun) Carriers and the Mk III, used as a smoke discharger for use by tanks. The 2in mortar remained in service until the late 1970s, when it was replaced by a modern 51mm weapon of advanced design. Typically, one single 2in mortar was allocated per rifle platoon headquarters and required a two-man crew, but it could just as easily be operated by one man in an emergency. For example, during fighting to capture the bridge at Bènouville over the Orne river on 6 June 1944, Sergeant 'Wagger' Thornton, who had landed by glider, engaged the German positions guarding the crossing using a 2in mortar. Despite the light weight of the bombs, the British attack was sufficient to force the Germans to withdraw and a bridge was captured intact which would later be needed to allow the Allies to move westwards out of the bridgehead. This was the type of role for which mortars were intended: to provide fire support. Two years previously, during the large-scale raid against German defences at Dieppe on the French coast during Operation Jubilee, a force of twenty men landing at Belleville-sur-Mer used their rifles and a 2in mortar to engage a gun battery known as the 'Goebbels' battery, which included three guns of 170mm calibre and four of 105mm calibre. So intense was their attack that the battery, despite its strength, was not able to direct any effective fire against the main Allied assembly on the beaches.

Private N.L. Francis was serving with 'C' Company, 4th Battalion, the Dorset Regiment, during Operation Market Garden in September 1944 and operated as the No. 2 on a 2in mortar. He recalled carrying a heavy load which included 'pack, pouches, gas mask, gas cape, entrenching tool, water bottle, bayonet, steel helmet, rifle, two bandoliers of 100 rounds of .303 ammo., Spare Bren gun mags, grenades and a pick or shovel down the back of your pack'. He continued: 'I got picked as No. 2 on 2in mortars and had two containers of bombs to carry, not much compared to WWI lads but enough to stop record attempts on a four-minute mile.' Major Mott served with

Figure 42. Two versions of the British Army 2in mortar. The weapon on the right is fitted with the so-called Palestine baseplate.

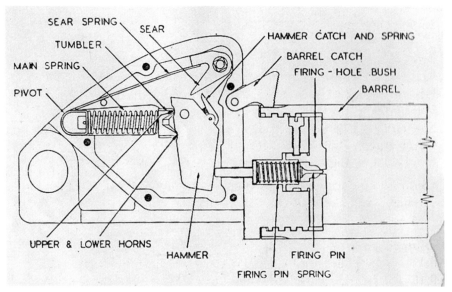

Figure 43. Diagram showing firing mechanism of the 2in mortar.

B Company of the 1st Hampshire Regiment. Landing on D-Day, he remembered the heavy load each man carried including '50 rounds of small arms ammunition, Bren magazines, four grenades, 2in mortar bombs as well as special equipment'.

An experimental grappling type projectile used to carry a rope either for climbing or to clear away obstacles was developed but it is not entirely clear whether it entered service because very few references to it can be found. An example of the actual projectile is kept in the weapon collection of the British Army's Small Arms School at Warminster in Wiltshire. The body has six blades with deep grooves cut into them, obviously designed to catch onto barbed wire. The base end with the propellant charge is weighted so that the projectile tips end-over in flight to allow the heaviest end of the body with the notched blades to engage. A long projector fitted with a pointed tip is fitted with a light rope, and this would dangle down free. Presumably, heavier-duty rope could be fitted for climbing or used to snag into barbed wire obstacles and the troops could then pull on the rope to clear a passage through the wire. This would be possible with five or six such projectiles fired into the wire. In view of the lack of documented evidence of its use, it would seem this was an experimental device and never entered service proper. Certainly it would have been useful to specialist units such as the Royal Engineers, whose task it was to clear minefields and other obstacles, and they may have had a use for it.

2in mortar bombs		
Type	*Colour*	*Weight*
HE	OD body, red band	2.25lbs
WP smoke	Dark green body	2.25lbs
FM smoke*	Dark green body	2lbs
Illumination	Drab khaki (light OD) body	1lb
Multi-star†	Light stone (grey) body	1lb (white 2lbs)

* Burning-type white smoke, titanium tetrachloride (FM).
† Available in white, red, green and red/green mixed.

Variations on the British 2in mortar

Mk I   = Introduced in 1918 and declared obsolete in 1919.
Mk II   = The first model introduced in 1938 with a large baseplate.
Mk II*   = The 1938 version intended for use with the Universal (Bren Gun) Carrier.
Mk II**   = A second version for use with the Universal Carrier.
Mk II***   = Version for use by infantry at platoon level and fitted with large baseplate.
Mk III   = Version used as smoke launcher for tanks.
Mk IV   = Limited production run and did not enter service.
Mk V   = Not manufactured.
Mk VI   = Not manufactured.
Mk VII   = For use on Universal Carriers.
Mk VII*   = For use by airborne forces, having shorter barrel and baseplate replaced with a spade-like plate.
Mk VII**   = Infantry use with long barrel and spade-like baseplate.
Mk VII*   = Same as above; not to be confused with airborne version with same designation.
Mk VIIA   = Indian Army model.
Mk VIII   = Another short-barrelled version for the airborne forces.

The 3in Mortar

The British 3in mortar was actually 3.2in calibre when adopted by the British Army, but like the 81mm mortars are sometimes rounded down to 80mm or 8cm, so the calibre was rounded down to 3in for simplification. The actual diameter of the barrel being 3.209in equates to 81.5mm, which meant that

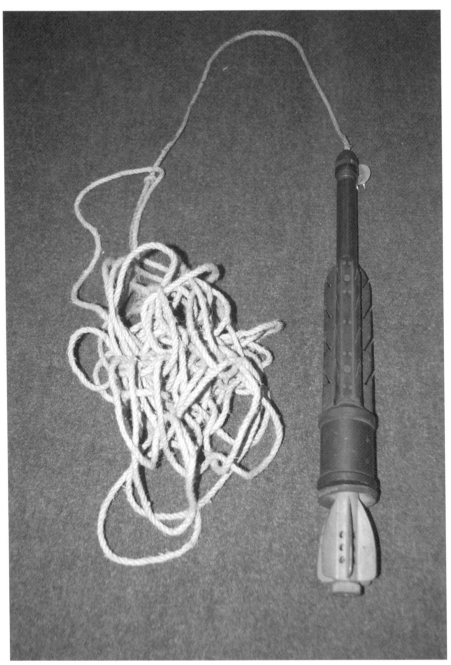

Figure 44. Experimental grappling-type projectile for the British Army 2in mortar. It was intended to clear barbed wire obstacles.

in an emergency it was possible to fire captured German and Italian 81mm rounds. It was already in service when war broke out and crews were well trained in its use. Like most designs of mortars it was a simple weapon, yet extremely versatile and could be modified to suit various situations, something which was frequently exploited as the war progressed. For example, the Canadian Army developed a version which used an extended barrel with a length of 81in designed to produce longer ranges. This was a massive increase of 30in in length over the original barrel length of 51in and increased the weight in action by another 32lbs. It was hoped that that the new development would provide a considerable increase in range, but firing trials showed the best that could be achieved was only an extra 300 yards. This was disproportionate to any advantages offered and the idea was soon dropped. The Australian Army decided to go the other way and reduce the original barrel by 40 per cent to produce one measuring only 30in in length. It was hoped this shortened barrel would be better to use in the jungle conditions of the Far East. There were side effects attached to this reduction in the length of the barrel, which resulted in accuracy being compromised. To compensate for this, special fast-burning cartridges were developed. Despite this, it still proved useful for jungle fighting but the version was only ever used by Australian forces.

The standard version of the 3in mortar weighed 126lbs in action, a load which was broken down into the three component parts of a standard mortar, comprising the barrel (44lbs), bipod (45lbs) and the baseplate (37lbs). Each of these parts was light enough for the three-man crew to carry as separate loads, but it meant that additional supplies of ammunition had to be carried by other infantrymen in the unit. Another method of transporting the 3in mortar was to place it in a Universal (Bren Gun) Carrier, which could also move significant stocks of ammunition, and the weapon would be deployed when the destination had been reached. The barrel could be elevated between 45 and 80 degrees, with 5.5 degrees traverse either side of centreline on the bipod. If

3in mortar bombs		
Type	*Colour*	*Weight*
HE Mk 6	OD body, red band	10lbs
WP smoke	dark green body	10lbs
FM smoke	dark green body	10lbs
Illumination	drab khaki (light OD) body	10lbs

Figure 45. French troops using the British 3in mortar after Dunkirk, 1940. Note the large springs fitted to the barrel to help diffuse recoil forces.

a greater traverse was required, the crew could pick up the weapon and simply move it around on its own axis. The crew could fire ten rounds per minute, with the original Mk 1 version achieving a range of 1,600 yards with a 10lbs HE bomb. A heavier barrel was developed and a stronger baseplate produced the Mk II, which could achieve ranges of 2,800 yards using the same ammunition, but the bombs were fitted with extra increment charges for the increased range.

As with all mortar designs of this calibre, the breech end of the barrel rested on the baseplate and the muzzle end was supported by a barrel clamp on the bipod. This did not secure the barrel in place; it was left to slide though it freely. Two large springs were attached to the barrel and the bipod so that when fired the recoil forces were directed through the baseplate and the barrel returned to its position by the springs. This meant that the bipod did not jump as much due to recoil forces and accuracy was maintained. To provide fire support for the mortar platoon within an infantry battalion's support company, a total of six 3in mortars was the normal compliment. The forty-six-man platoon was divided into six mortar detachments of five men each, with a sixteen-man headquarters. The large headquarters possessed four trucks to transport tools and spares along with a significant amount of ammunition. Other versions were also trialled but like the Canadian version with its extended barrel, the weight was greater than any advantages in range. Development of further designs of the 3in mortar ended with the Mk V, which was fitted with a modified firing pin to permit the firing of captured stocks of German and Italian ammunition. Considering that the original version of the weapon was capable of this practice, it seems like a waste of resources. This weapon, like many other successful types, continued in service with various armies around the world for many years after the war.

The 60mm M2

Armaments factories in America increased their output of weapons and very soon the army was making up its deficiencies from the early part of the war. It kept things simple and, like its British counterpart, concentrated on just two calibres of mortar designs for its infantry, which also proved satisfactory for other units including Rangers and Airborne. The lighter weapon, the 60mm M2 with an M2 adjustable bipod mount, was a slightly modified version of the French company of Edgar Brandt, built under licence. A series of firing trials were concluded in the late 1930s and by January 1940, the first order for 1,500 weapons had been placed. Unusual for a light mortar design, the M2 had three component parts, more normally associated with heavier weapons.

60mm mortar ammunition		
Type	*Colour*	*Weight*
HE M49A2	OD body, yellow markings*	2.94lbs
WP smoke M302	Grey body, yellow band and markings	4.02lbs
Illumination M83	Grey body, black markings	3.72lbs
Practice M50A2	Light blue body, white markings	2.94lbs
Drill M69 (inert)	Black body, white markings	2.94lbs

*Prior to 1943 HE had a yellow body with black markings.

The barrel weighed 12.8lbs, with the bipod and baseplate weighing 16.4lbs and 12.8lbs respectively to give it a weight in action of 42lbs. In keeping with the heavier weapons, the M2 was fitted with a fixed firing pin which allowed the crew to achieve a very high rate of fire in the order of eighteen rounds per minute. All of this meant that the M2 was more powerful than many other comparable weapons in this range. For example, with a maximum range of nearly 2,000 yards, it bettered the French 60mm *Modele* 35 by 200 yards and outdistanced the 70mm Japanese Type 11 by 300 yards.

A crew of three men served the M2 mortar, with extra ammunition being carried by other members of the platoon in addition to their standard personal combat equipment. It proved a popular weapon with troops in all theatres of operations. A typical US Army rifle company weapons platoon had a mortar section with three five-man squads, each with one mortar. The US Marine Corps used essentially the same organisation, but with only two mortar squads until April 1943, when it was increased to three squads. In May 1944, the rifle company weapons platoon was converted to a pure machine gun platoon and the 60mm mortar section was reassigned to the company headquarters. America also supplied the M2 to their allies, such as French troops for D-Day, and over 4,000 were supplied to the Chinese Nationalist Army fighting the Japanese. The Chinese were so impressed by the design that they produced their own copy of the weapon, which was known as the Type 31 mortar. Each type of ammunition for use with the M2 had its own designation, such as the HE bomb that weighed 2.94lbs and was termed the M49A2. The illuminating round was the M83 and this produced an intense luminosity equal to 100,000 candle-power for 25 seconds as it descended on a small parachute. With several M2 mortars firing in conjunction, it was possible to illuminate a wide area sufficiently for machine-gunners to engage targets, and for this reason it

Figure 46. Sight unit on US M2 60mm mortar. (*SASC*)

was also issued to anti-tank gun units for use against armoured attacks in low light conditions. The M302 WP smoke was introduced late in the war. Total wartime production of ammunition for the M2 mortar is believed to have been in the order of 51,756,000 rounds.

When the airborne forces parachuted into Normandy, four men jumped with a mortar and three further members of the mortar crews jumped with additional ammunition carried in M6 bags. A total of eighty rounds of ammunition were dropped for each weapon and the cart could be used to transport fifty-four rounds. A lightweight version of the M2 mortar, known as the T18E6, was developed for the airborne units, and was standardised as the M19 after the war. The baseplate was reduced to a curved spade-type base and the bipod was discarded. It was also fitted with a selectable trigger, allowing the weapon to be operated using either the trigger or normal drop fire. Along with other modifications, the weight was reduced to only 20lbs as opposed to the original 42lbs, but only fixed charges could be used with the ammunition and the range was reduced to 816 yards. Even then, this would have proved more than sufficient for use with airborne units, who could have made good use of it had it been available for actions such as the Normandy campaign and Operation Market Garden, where they had to fend for themselves until relieved by heavier armoured units. The T18E6 could also be mounted on the standard bipod and baseplate, and it is understood to have been used in limited service during the fighting on Okinawa between April and June 1945.

As well as practice and drill rounds being used for training purposes, a special firing mechanism was also developed for the M2 mortar. When screwed into the breech-end of the barrel, this device looked identical to the real section used in combat. However, it was fitted with a firing lever in place of the fixed pin on the actual live firing weapon to prevent accidental firing during training exercises.

The 81mm M1

The American 81mm M1 mortar on the M1 mount was another copy of a French Brandt-designed weapon, built under licence. It had been considered for issue to the American Army as early as the 1930s and the US War Department was more than satisfied with its performance during initial field trials. The only problem with the weapon in their opinion lay in the French-designed ammunition, which was considered too complicated because of the folding fin design of the bombs which they believed could become either bent on firing or fail to deploy. In March 1940, the American Army had developed the M56 round of ammunition, which was of more conventional design and

81mm mortar ammunition		
Type	**Colour**	**Weight**
Light HE M43A1	OD body, yellow markings*	7.05lbs
Heavy HE M45	OD body, yellow markings*	15.05lbs
Heavy HE M56†	OD body, yellow markings*	10.77lbs
WP smoke M57‡	Grey body, yellow band and markings	11.59lbs
Tear gas M57	Grey body, red band and markings	11.36lbs
Illumination M301	Grey body, white markings	10.5lbs
Practice M43 & M44	Light blue body, white markings	7.05lbs
Drill M68 (inert)	Black body, white markings	7.0lbs

*Prior to 1943 HE had a yellow body with black markings.
† M56 more widely used than M45.
‡ Also filled with FS, sulphur trioxide and chlorsulfonic acid.

handled in the same way as the M45 it replaced. A complete range of ammunition types was developed for the M1, and between July 1940 and August 1945 a stock of 81mm ammunition in the order of 37,043,000 was produced and the mortar was used in all theatres of the war. The US Army organised infantry battalions to incorporate a heavy weapons company which was structured to incorporate an 81mm mortar platoon with three sections. Each of these section had two squads with one mortar, each manned by seven men. The weapons company of a US Marine Corps infantry battalion was organised along similar lines, but with only two sections with two squads each. In May 1944, the weapons company was eliminated; the mortar platoon was reassigned to the battalion headquarters company and the machine guns to the new rifle company machine gun platoon.

Training rounds were used to familiarise crews with operational use of the mortar, but practice bombs could only provide so much experience and troops had to fire live ammunition to gain confidence. This was done on training grounds both in America and Britain as troops arrived in preparation for the build-up for D-Day. One area designated as a live firing training ground for American troops was Saunton Sands in North Devon. Here, concrete replicas of German bunker defences were constructed and assault landings practised. It was inevitable that some accidents should occur during these exercises and some were fatal, such as the incident witnessed by Wesley Ross, who served with the 146th Engineer Regiment (Combat) remembered seeing 'two men killed when an 81mm mortar bomb blew up in the tube'. There was always a possibility that something like this could happen, even in battle, but such an event was still a very low risk factor in the overall scheme of things when taking into account the number of bombs fired in a battle. When parachuting into Normandy, each M1 mortar was dropped in a bundle carried by three men along with a total of fifty-four rounds of ammunition for each weapon. The carts used to transport the ammunition could carry up to thirty rounds and four men each carried an extra six rounds. For every eight rounds of HE there were two white phosphorous smoke bombs. An alternative distribution rate for the ammunition was four men each carrying five rounds and the fifth man carried only four rounds. This still represented a considerable weight for each man to carry in addition to his parachute, personal kit and weapon.

The HE M43 bomb weighed 6.8lbs and the M1 could fire this out to a maximum of 3,290 yards by using up to six secondary charges in addition to the primary propellant charge. The same principle applied to the heavier HE M56, which weighed 10.6lbs and had a range of 2,558 yards, whilst the M57 smoke bomb could be fired out to 2,470 yards to screen troop movements. The heavy HE was expended at higher rates than anticipated, as it proved more effective against sand-bagged positions such as machine-gun posts and mortar sites, compared to the light HE. The barrel was 49.5in in length and weighed 44.5lbs. The bipod and baseplate added a further 46.5lbs and 45lbs respectively to give the M1 a combat weight of 136lbs. To ease the burden of transportation of such a load, the M6A1 Hand Cart was designed, issued at the rate of four carts per company, and even special harnesses for mules were developed for transporting the mortar by pack animal. These two types of equipment proved to be extremely useful during the hard fighting in the mountainous terrain of the Italian campaign from 1944 onwards.

The Soviet Red Army

The Soviet Red Army used the greatest numbers of mortars during the war, both in terms of numbers of weapons and separate designs. They successfully deployed mixed batteries of the heavier calibre weapons along with units of conventional tubed artillery and lighter calibres to provide local fire support in the event of an attack by infantry. Very early on in the war, the Soviets realised the usefulness of the rapidity with which mortars could be brought into action and the level of firepower they could deliver on to a target within a short space of time. Before the war the Soviets were already using a wide number of mortar designs, but during the war against Germany and its allies they standardised on four main types of 50mm mortars, these being the RM38, RM39, RM40 and RM41, along with four types of 82mm calibre mortars, these being

Captured Soviet mortars used by the Germans	
Soviet Designation	*German Designation*
50-RM 38 5cm	GrW 205/1(r)
50-RM 39 5cm	GrW 201/2(r)
50-RM 40 5cm	GrW 205/3(r)
50-RM 41 5cm	GrW 200(r)
82-BM 36 8.2cm	GrW 274/1(r)
82-BM 37 8.2cm	GrW 274/2(r)
82-BM 41 8.2cm	GrW 274/3(r)
107-PBHM 38 10.7cm	GebGrW 328(r)
120-HM 38 12cm	GrW 378(r)

PM36, BM37, BM41 and BM43 82mm. These weapons were joined by the PBHM38 mountain mortar and the HM 38 of 120mm calibre. It was this last calibre of weapon which the Germans would go on to use against the Soviets themselves, having re-designated it and even copying it as the *Schwere* 12cm *Granatewerfer* 42. In addition to this large calibre mortar, the Germans also captured vast quantities of other designs of mortars, some of ageing pre-war design, ranging from 5cm to 8.2cm, and in keeping with their usual practice concerning captured foreign weapons were given a standard German Army nomenclature for identification purposes.

In the 1939–40 Winter War against Finland, a typical Soviet Red Army rifle division comprised of three rifle regiments, each of three battalions. Each of these rifle battalions, in turn, was formed into three rifle companies of 140 men with a single support company equipped with machine guns and mortars. By way of reply, the Finnish Army was equipped with 81mm mortars for their support company units. The rifle battalion structure of the Soviet Army was one which would remain in place until the end of the war. By 1942, only one year after the German invasion, the Soviets had reformed and were fighting back with divisions some of which had a total 'gun strength' counting 160 mortars. Specialised mortar brigades were formed within divisions, for which purpose they were equipped with at least 100 mortars of 120mm calibre. It was this kind of unit which would eventually fight its way through to Berlin in 1945, along with other conventional artillery, to take the fight right into the German capital. The Soviets did have a number of other designs in service, but it was those four types which served as the backbone of the country's infantry units. Contrary to popular conception, Soviet 82mm mortar ammunition could be fired in any 81mm mortar. The Germans, though, thought they could not and while they fired their ammunition in captured Soviet mortars, they did not fire Soviet ammunition in their mortars.

The 50mm mortars were allocated two per rifle company in a small mortar section under the company headquarters. These sections were eliminated by 1943. The rifle battalion's sixty-one-man mortar company had three platoons with three 82mm mortars each. The rifle regiment possessed a seventy-man mortar battery with two platoons. Each platoon had two sections of two 120mm mortars for a total of eight in the battery.

The 50mm RM Series: RM38; RM39; RM40 and RM41

Despite the numbers in service with the Red Army, the range of 50mm calibre light mortars never seemed to have been widely popular with the Soviet forces; the troops preferred instead to have something with greater hitting power. The range of 50mm mortars were designated *Rotney Minomyot obr.* (Company Mortar) for use at platoon level within an infantry company, and included four weapons. They all fired the same ammunition, with the HE bomb weighing around 1.87lbs and able to be fired out to ranges of around 880 yards. The first mortar of 50mm calibre entered service with the Soviet Army in 1938 and was augmented in quick succession by a series of other designs, which concluded with the RM41 Model, but together they satisfied the army's requirements at the time. The RM38 weighed 26.6lbs in action, with a barrel length of around 30in, and was fitted with a baseplate and bipod to permit firing at angles of elevation between 30 and 60 degrees. It had a firing rate of up to thirty rounds per minute, which could also be achieved by other RM designs. The barrel of the RM39 was about the same length but it could be adjusted to firing angles between 46 and 85 degrees and was a much heavier weapon at 35lbs. The RM40 was a more compact weapon with a barrel length of only 24.8in, which could be angled to fire between 45 and 75 degrees, and weighing 24.8lbs in action. The last weapon in the range was the RM41, which was different from earlier styles, such as the RM40, replacing the separate bipod and buffer units by hinging the barrel to the baseplate and fitting the weapon with a venting system to exhaust excessive propelling

50mm mortar ammunition		
Type	*Colour*	*Weight*
HE	OD	1.875lbs

gases through an outlet at the base of the barrel. The barrel of the RM41 was 24in in length, being angled to fire through 45 to 75 degrees and weighing 22lbs in action. At the time of the German invasion in June 1941, a Soviet Red Army infantry division had eighty-four *Rotney Minomyot* 50mm mortars in its formation. However, by December 1944, when the Red Army was poised to invade Germany, there were no 50mm in service with any infantry division, the weapons of 82mm calibre having replaced the lighter weapons. This change in favour of heavier calibres reflected the evolving tactics of the Soviet Army as it moved from the defensive actions of 1941 and 1942 to the heavy offensive actions from 1943 onwards, which required heavier firepower which could only be delivered by heavier calibre weapons.

The 82mm BM Series: BM36; BM37; BM41 and BM43

The Soviet Army had a large number of 82mm mortars in service before the invasion of 1941, all of which were with front line units, and as a consequence vast stocks of these weapons were captured during the rapid German advance. These were the BM range or *Batalyonny Minomyot* (Battalion Mortars) and would include four types of weapon. The first was the BM36, which had a barrel length of 4.3ft which could be elevated to fire between angles of 45 and 85 degrees and weighed 136.7lbs in action. It could fire bombs at the rate of twenty-five to thirty rounds per minute and the HE bomb weighing 7.5lbs had a range of 3,300 yards. In 1937, this was followed into service by the BM37, which some sources credit as being designed by N. Dorovlyov whilst others claim it was designed by B.I. Sazayrin. The barrel was the same length as the BM36 and all other factors, including the range and type of ammunition fired, were identical. The major difference was that the BM37 weighed 126.3lbs in action. As more weapons were destroyed in the fighting a new design, the BM41, was introduced to replace those weapons lost. It was immediately put into mass production, with many thousands coming direct from the factories to be issued to front line units. The barrel of the BM41 remained 4.3ft in length but it weighed around 100lbs in action. It was fitted with a pair of removable wheels on a short stubby, allowing it to be towed by a light vehicle, and in this mode it weighed 143lbs. It could be angled to fire

82mm mortar ammunition		
Type	*Colour*	*Weight*
HE	OD	7.4 and 7.5lbs

Figure 47. Soviet Red Army troops with an 82mm mortar, probably the BM41.

between 45 and 85 degrees of elevation and the range of 3,300 yards with the 7.4lbs bomb remained the same, as did the rate of fire. The BM41 could be traversed 3 degrees either side of the centre line on the bipod, but for greater traverse the crew simply moved the whole weapon round to engage new target areas. In 1943, a modified version of the BM41 was introduced into service as the BM43. This featured an improved bipod design and the incorporation of permanent wheels for ease of moving the weapon quickly in one piece, ready to fire, by the infantrymen, who could pull it into action in an emergency. The medium calibre 82mm mortars were to remain the mainstay fire support weapons of the rifle divisions of the Soviet Army during the war, and numbers in service increased. For example, in late 1941, a typical rifle division had eighty-four mortars of 82mm calibre in service within its structure, but this figure rose to ninety-eight, an increase of almost 12 per cent, by the end of the war. This not only significantly increased the firepower but also more than filled any gaps left in the support units when the 50mm was taken out of front line service.

The Soviet Red Army had heavier mortars in service with infantry, such as the PBHM38, which had been designed in 1938 for use with specialist mountain troops. This weapon had a calibre of 107mm and was referred to as being a 'mountain mortar'. It had a barrel length of just over 51in and weighed 376lbs in action. It was in service from 1942 and fired a HE bomb weighing 17.6lbs out to a maximum range of over 6,900 yards. The Germans used captured stocks of this weapon, which they designated as the 10.7cm *Gebirgsgranatwerfer* 328(r). It could be transported on a light limber towed by a truck, or even pulled by horses. Alternatively, it could be broken down as pack loads for individual animals, which was ideal for use in the mountains. Firing was by the usual 'drop' method but a trigger mechanism was also fitted as an alternative in an emergency. The last mortar type to be used by infantry units of the Red Army was the HM38, sometimes termed as PM38 (*Polkovoi Minomyot* or Regimental Mortar), which had been designed in 1938. The HM38 mortar had barrel length of 6.13ft and could be elevated between 45 and 80 degrees. It weighed 617lbs in action and was towed on a two-wheeled carriage. The HE bomb weighed more than 35lbs and could be fired out to ranges of over 6,500 yards. By comparison, the version developed for the German Army had a barrel length of 6.1ft and could be elevated between 45 and 84 degrees. It weighed over 628lbs in action and could fire a HE bomb weighing 34.4lbs out to ranges of more than 6,600 yards. The German Army version may have been only slightly heavier and the design produced only a marginally better range, but it was to remain a standard design in service and was even supplied to allies such as the Finnish Army, which also produced their own version known as the 120krh/38.

The German Army

According to sources it is understood that in 1939 the newly reformed and recently re-armed German Army was fielding perhaps about the same number of 81mm mortars as the French Army, which is to say approximately 8,000 weapons. By 1940, the regimental firepower of the German Army was structured to comprise eighteen mortars of 81mm calibre and a further twenty-seven of 50mm calibre. During the Polish and French campaigns of 1939 and 1940, heavy mortar platoons were structured within the German Army. These had a strength of three sections, each with nineteen men and two 81mm mortars. Later in the war, on the Soviet Front from 1941 onwards, the Germans added to their mortars with large numbers of captured Soviet weapons, including hundreds of mortars of 120mm calibre. So many of this particular weapon were captured that it became viable for the Germans to utilise them against their former owners and even create special Support Companies for infantry regiments, which were equipped with four 120mm mortars and six 81mm mortars. By 1944, fortunes had been reversed with retreats on all fronts, and the German Army found itself suffering from severe manpower shortages, especially in the wake of the massive defeats on the Eastern Front. These losses, along with those in North Africa, Normandy, Greece and Italy, forced them to reorganise and restructure the infantry divisions, reducing the manpower levels by more than 5,000 troops to 12,772 compared to 17,734 troops in the divisions of 1939. The number of infantry regiments in these newly-structured divisions remained at three but the numbers of battalions to each was reduced to two, with four 120mm mortars and six 8cm mortars per regiment. The light 5cm mortar was taken out of service in 1943 along with some of the captured weapon stocks, which after three years of war were now obsolete due to wear and tear or lack of proper ammunition.

The 5cm *Granatwerfer* 36

From the very start of the war, the German Army placed a great deal of store in mortars of various calibres and deployed them to every theatre of war, from North Africa to the Balkans and north-west Europe. The lightest calibre mortar produced expressly for the German Army was the 5cm *leichte Granatwerfer* 36 (leGrW36), which had a weight in action of 30.9lbs, considerably heavier than anything used by the Allied armies. Despite this it fired a HE bomb of just under 2lbs in weight, which was less than the weight of the bombs fired by the British 2in mortar or the Japanese Model Type 98 of comparable calibre. At the start of the war, the leGrW36 was standard equipment with every platoon within an infantry regiment of the German Army and required three men for its operation. It was used at company level to provide the company commander with immediate fire support to platoons and sections. At the time of Operation Barbarossa, the German invasion of the Soviet Union in June 1941, there were eighty-four of these mortars in service with each division. The crew between them carried forty-five rounds of ammunition ready to use, and with a firing rate of forty rounds per minute this gave them just over one minute in action. However, these supplies would not have been expended so quickly and targets would have been engaged selectively in order to conserve ammunition and provide the maximum fire support to those

5cm mortar projectiles		
Type	*Colour*	*Weight*
HE Wgr36	maroon body	1.98lbs
FS smoke Wgr36Nb	maroon body, white Nb	1.98lbs

areas where it was most needed. The mortar had a barrel length of 19.3in and could fire at angles of elevation between 45 and 90 degrees, with a maximum range of 550 yards with a HE bomb.

The ammunition for the 5cm mortar was termed *Werfergranate* 36 and was made with a cast-steel casing. It was fitted with the *Werfergranatzunder* 38 fuse, which also had 'graze' action. For safety purposes the fuse only became armed some 60 yards out from the muzzle after firing. The graze action on the fuse meant that if the bomb were to descend through trees, there was a high possibility that it would brush against branches which would set off the detonation train of the fuse to produce an air burst action. This type of fuse was fairly common at the time and other armies also had versions incorporating similar actions for the same effect. The ammunition for the leGrW36 was carried in pressed steel containers, each holding ten rounds ready to use.

It has been argued that the 5cm mortar of the German Army was actually over-engineered and fabricated from the best quality metals. This is supported by the comments made in a report following an examination made on captured examples which were also test fired in 1941, by the British Army which stated that is was 'well-constructed and easy to operate, but the degree of accuracy is unnecessarily high'. The barrel was attached to the baseplate by means of a locking pin, which allowed it to be moved through its arc of firing independently of the baseplate's angle to the ground. The first examples of the mortar had been issued during the re-armament programme in the mid-1930s and these were fitted with a collimating sight. It was later decided to dispense with this, as on the British 2in mortar. Like that weapon, it was left to the experience of the firer to judge the angle of the barrel when in use. The leGrW36 could be broken down into two parts for carrying by the crew. This was achieved by removing the locking pin and disconnecting the elevating mechanism so that the barrel and baseplate became two separate loads. The 5cm mortar was used by *Gebirgsjäger* (Mountain Troops) during operations where they were engaged in anti-partisan actions among the peaks in regions such as Crete, Greece and Yugoslavia. The Germans nicknamed it the *Zigeuner-Artillerie* (Gypsy Artillery) because the weapon was quick and easy to move from one location to another on the battlefield. Despite its usefulness in such terrain, the Germans came to realise the weapons were too heavy, too expensive and too complex for the limited downrange effect they provided and eventually decided to follow the move of the Soviet Red Army concerning the use of 5cm mortars, and by 1943 the GrW34 had been withdrawn from service. Production of ammunition was halted but it remained in service with some of the more remote garrisons such as the Channel Islands, where stocks of ammunition were unused, except for a few rounds expended for training purposes.

The *Gebirgsjager* were specialist infantry with a typical division having 14,131 troops of all ranks, 3,506 mules and horses along with more than 550 bicycles for the 'Cyclist Battalion' within its organisational structure. The bicycles were a cheap and efficient method of moving a battalion over distances without having to rely on motor transport. Some of the bicycles were fitted with brackets at various points, such as the crossbar and handlebars, to permit heavy weapons including machine guns and the 5cm mortar to be carried. Ammunition could be carried in panniers on the rear or in special saddlebags. The bicycle battalion of the *Gebirgasjager* had six 5cm mortars and three 8cm mortars, while the two infantry regiments combined had a total of fifty-four GrW36 and thirty-six GrW34.

The Germans occupied the British Channel Islands in July 1940 after the surrender of France, and the islands would remain in German hands for almost five years. During that time the three main islands in the archipelago, Jersey, Guernsey and Alderney, were fortified out of all proportion to their actual strategic value. The garrison spread across the islands increased to an eventual strength of around 46,000, including some 15,000 forced labourers who were taken to the islands to build the defences which encircled each of them. Although there were elements of the *Luftwaffe* and *Kriegsmarine*, which provided anti-aircraft batteries and coastal artillery units, the main bulk of the force was provided by the infantry. Initially this had been 216th Division but this was replaced by 319th Division in May 1941. The infantry deployed all the usual standard weapons such as machine guns and flamethrowers, but they also prepared defensive emplacements for mortars, including the 5cm GrW36 and the heavier 8.1cm GrW34, of which there were about thirty-nine and forty-three respectively deployed to Jersey. One of these positions was located at La Crete Fort, built on a promontory between the harbour at Bonne Nuit Bay and Giffard Bay on Jersey's north coast. The site had been used as a defensive point since the seventeenth century and updated in the nineteenth century. The buildings were modified by the Germans to accept mortars and machine guns, and included a guardhouse that could be used by the garrison of twenty troops. The site overlooked the sea on three sides and from there the mortars could be traversed in all directions with unimpeded views to engage any would-be assault force. Another mortar-firing position was built at St Aubin's Fort, which dates from the sixteenth century and lies at the western end of St Aubin's Bay on the south coast of the island. The special emplacement had a concrete platform from which the crew could traverse the mortar to engage targets at all angles of approach.

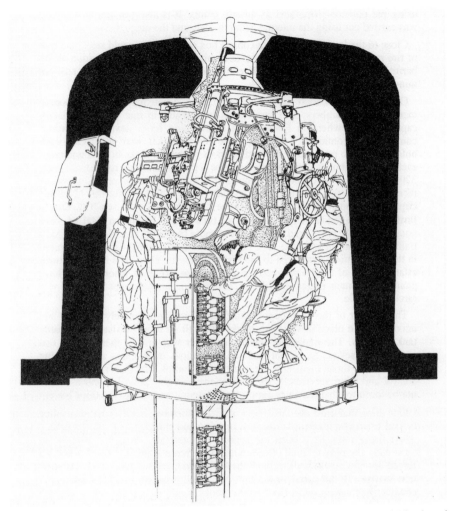

Figure 48. Diagram of German M19 5cm automatic mortar as sited in the Channel Islands and at points on the Atlantic Wall.

Apart from the two standard types of service mortars, the occupying forces also bolstered the defences with captured weapons which included a range of French and Soviet mortars. On Jersey this included at least twenty-eight French 50mm mortars which were given the German Army service nomenclature of 5cm GrW210(f) and are understood to have been taken from the defences of the Maginot Line. Soviet weapons included eleven 52-PM 37 mortars known in German service as 5.2cm GrW205(r) and a few 82-PM 36 which the Germans referred to as the 8.2cm GrW274(r). These began arriving on the island in late 1941 and 1942. One of the units recorded as using these weapons on Jersey was the 4th Battalion, 582nd Infantry Regiment, which included the 643th Battalion, the *Russiskaya Osvoboditelnaya Armiya* (Russian Liberation Army). Soviet troops volunteering to serve with German units were posted at low key defensive positions along the Atlantic Wall in those areas where it was considered unlikely that an Allied invasion would come. One of these areas was the Channel Islands and, as such, most of the Soviet troops sent here would have been trained in the use of Soviet weaponry.

The islands had been under occupation for more than fifteen months when Hitler issued a directive on 20 October 1941 which ordered that permanent defences be prepared on each of the islands. Directive No. 441760/41 for the 'Fortification and Defence of the British Channel Islands' was marked 'secret' and contained instructions that: 'Defence measures on the Channel Islands must guarantee that a British attack will be repulsed before reaching the islands irrespective of whether the attacks are by air, by sea, or a combination of both. It must be taken into account that the enemy may use bad weather for a surprise attack. Immediate steps to strengthen the defence measures have already been ordered.' The campaign into the Soviet Union was only three months old and going very much in Germany's favour, and one would have thought this would be Hitler's main priority, but in this directive we see that he is concerning himself with a minor aspect in the overall conduct of the war. This concern has led historians to conclude that he became obsessed with holding on to these British possessions at whatever cost. The war was also going in favour of Rommel's *Afrika Korps* in North Africa, where the British and Commonwealth forces had been pushed back to Egypt. Yet, here he was devoting time to draft directives concerning occupied territories of little importance other than for propaganda.

The islands had already been subjected to several raids by British commando units, conducted mainly against Guernsey, with the aim of establishing the strength of the garrison. By interrogating prisoners, the British were able to build up a picture of what was happening by 1941 and gauge morale of both the civilian population and the military government. These raids achieved more through their irritating effect by keeping the Germans on a state of alert rather than any genuine military success. However, they did create a reason in Hitler's mind to continue defending the islands. His October 1941 directive continued by stating: 'For the permanent fortification of the Channel Islands, which must be pressed forward energetically in order to create an impregnable fortress, I therefore order the following ...' The document continues by outlining the responsibilities of the *Luftwaffe* and *Kriegsmarine* on the islands, but emphasis is given in particular to the role and

Figure 49. German M19 5cm automatic mortar in action, being loaded with trays of ammunition.

duties of the army. One of the paragraphs of the directive states that: 'For the Army the most important constructions are close-meshed flanking installations spacious enough to contain guns with a calibre sufficient to penetrate armour 100mm thick, and for defence against tanks which may be landed from barges.' Recommendations for storage of ammunition are all outlined and how such construction work, in keeping with installations along the Atlantic Wall, would be built using 'foreign workers, especially Russians, Spaniards, but also Frenchmen'. These were slave labourers under the direction of Organisation Todt. The Atlantic Wall defences they built would eventually stretch for over 1,600 miles, from Norway to the Spanish border. It absorbed 18.6 million cubic yards of concrete and almost 1.2 million tons of steel to create thousands of emplacements, from the smallest to the largest, which housed long-range guns which could fire shells more than 30 miles across the English Channel to bombard the town and harbour facilities at Dover in Kent.

There were many different types of installations built, into which were mounted weapons of all types from machine guns to massive pieces of artillery capable of firing shells many miles. One design mounted automatic mortars in armoured cupolas, an idea which had occurred to military planners as early as 1934, and the Dusseldorf-based company of Rheinmettall-Borsig was awarded a contract to develop such a design. At the time the French Maginot Line was at an advanced stage of building and some of the defences incorporated mortars mounted in steel cupolas, and this idea may have inspired the German planners. The French opted to equip their cupolas with either 50mm or 81mm mortars, while the Germans decided on just mortars of 5cm calibre based on the GrW36 weapon, which, by coincidence, was also manufactured by Rheinmettall-Borsig. One theory put forward as to why the lighter calibre was chosen is because it would not have placed too much of a strain on production and supply in the way a larger calibre would have. Armament factories in Germany could have coped with extra production and certainly those armaments factories in occupied countries, such as France, could have added to the supply. The truth is probably due to size and weight of the ammunition, which could be handled by infantrymen instead of requiring specialist handling equipment. The M19 automatic mortar fired standard bombs from a metal tray-like magazine which was pre-loaded with six bombs. The 5cm HE bomb weighed 1lb 15.5oz, and together with the weight of the complete magazine each would be just over 12lbs. If this had been done with the GrW34 8cm HE bombs, each of which weighed 7lbs 8oz, the weight would have been over 45lbs and the magazine tray would have been much larger and heavier. Indeed, size was a major contributing factor and to put it in simple terms of logistics, more of the smaller 5cm bombs could be stored inside the bunker than the larger 8cm bombs.

Rheinmettall-Borsig produced ten studies into developing a complete system for an automatic mortar before making a final choice which would become the M19 5cm *Maschinengranatwerfer*. The operational role of the system was to provide firepower to cover areas of 'dead ground' which could not otherwise be observed. This was usually an area on the coastline with steep cliffs, but that was not exclusive. Apart from firing the 5cm calibre mortar bomb, the weapon used in the M19 system was completely different to the standard GrW36 used by the infantry. Using standard dismountable mortars in such a defensive role would have only been a short-term solution and they would have needed to be removed periodically for service. Emplacing a weapon mounted in a specially-produced turret or cupola would provide a permanent position, ready to provide all-round 360-degree traverse and able to come into action at a moment's notice to cover all points of approach to the defensive site. Initially, these automatic mortars were intended for installation in the Westwall and the Eastwall, a defensive system also known as the Oder-Warthe-Bogen Line. This was built between 1938 and 1940 on the border between Germany and Poland. It covered a length of around 20 miles and included around 100 main defensive emplacements. After the successful campaigns in 1939 and 1940, it was decided not to install the weapons in these locations and instead they would be sited at intervals along the Atlantic Wall, which included several being built on the Channel Islands of Jersey, Guernsey and Alderney.

The M19 installation on the island of Jersey was built at Corbière Point at the western end of the island, which was turned into a strongpoint to defend the headland. From here its high rate of fire could be useful in engaging targets at close quarters and overlap with the firepower of machine guns and two 10.5cm field guns also sited at the point. The neighbouring island of Guernsey had four M19 automatic mortar installations, including one located at Hommet, overlooking Vazon Bay on the north-west coast, where its firepower could be integrated with that of machine guns, at least three pieces of artillery with 10.5cm calibre and a 4.7cm gun of Czechoslovakian origin. On Alderney there were two M19 strongpoints with other similar installations built along the much-vaunted Atlantic Wall, including three in Norway, nine in Holland, one in Belgium, twenty-two along the French coast and twenty along the Danish coastline, with four more planned but not built. For such a small weapon it absorbed a huge amount of resources in manpower to build the emplacement, with tons of concrete and steel in its construction. The sites of the M19 automatic mortars were out of all proportion compared to those built for heavier weaponry in defensive positions. The M19 mortar could fire HE bombs at a rate of between sixty and 120 rounds per minute, although the higher rate of fire was rarely used in order to minimise stresses and prevent the weapon from overheating. The crew could engage targets at ranges between 54 and 820 yards, which was closer than artillery could achieve, and together with support fire from other weapons such as machine gun, any infantry attack would have been met with fierce opposition. Indeed, one M19 position on the Eastwall held out for forty-eight hours when attacked by troops of the Red Army in early 1945.

The M19 weapons were mounted in steel cupolas which had an internal diameter of 6ft 6in to accommodate the three-man crew during firing. Initially, there were two main designs of cupola, the 34P8 and the 49P8, but it was a third type, the 424PO1, which became the most widely used with armour protection 250mm thick. The cupolas were mounted on specially-prepared bunkers designated '135', with concrete protection up to 11.5ft thick, and the '633', which was the most common design and the type used in the Channel Islands. The M19 bunkers were divided into several rooms including the firing room and had accommodation for up to sixteen men. Each bunker had its own independent generator to provide power to traverse the firing platform and cupola, but in the event of a power failure the weapon and cupola could be elevated and traversed by means of hand-operated wheels. The ammunition storage room had racks for thirty-four trays, each pre-loaded with six bombs, giving a total of 204 bombs ready to fire. Ammunition boxes containing ten bombs each to reload the spent trays were stored in this room, and it was the task of the crew members to reload these. In total an M19 bunker could have ammunition reserves of up to 3,944 bombs stored in readiness for use. These bunkers were equipped with field telephones and optical sight units such as the *Panzer-Rundblick-Zielfernrohr*, an armoured periscope with a magnification of × 5. Because it was an indirect fire weapon, the M19 had to be directed on to its targets and integrate its fire by overlapping with neighbouring weaponry.

The firing platform on which the mortar was mounted could be elevated when firing and lowered when not in use. The loaded ammunition trays were fed up to the platform by means of an elevator where the loader removed them and fed the trays into the left-hand side of the weapon's breech. As it fired, a mechanism moved the tray along to feed the next bomb into the weapon, and the process continued until the empty tray emerged on the right-hand side of the weapon, where a handler removed them and placed them in the descending elevator section. These were removed by another member of the crew and taken to the ammunition room, where they were reloaded ready for reuse. The M19 could fire the standard types of *Wurfgranate* 36 bombs, which these were fitted with colour-coded graduated propellant charges to be used according to the range required. The red charge was for use at ranges from 22 to 220 yards and the green charge was for ranges from 220 to 680 yards. There were training bombs which had no filling and could not be fired, which were really for familiarising crews with handling procedures of the weapon. There were two training systems developed to teach crews how to operate the M19; the first was the *Sonderanhangar* 101 mounted on a trailer and the other was the *Ubungsturm*, which replicated the complete cupola layout.

Preparing the weapon to fire, the operator lifted the barrel clear of the breech by means of a cam and lever mechanism which allowed a bomb on the loading tray to be aligned with the chamber. As the barrel was lowered over the bomb, the firing pin in the breech was activated to initiate the propellant charge. The recoil forces on firing unlocked the barrel a fraction of a second later and the cam mechanism lifted the barrel clear of the breech, and another bomb was loaded ready to fire as the tray was fed through.

As the Allied campaign to liberate Europe continued in the second half of 1944, the Germans had to rethink their defensive strategy. From September 1944 they renewed construction work on the Westwall to improve defences and began to install M19 systems at certain locations. The actual numbers of M19 systems built and turrets installed varies according to sources. Some, for example, state that perhaps seventy-three such installations were built in the Atlantic Wall. These did not cause the Allies any unnecessary problems and those along the Westwall were largely ineffective, while those in the Channel Islands never fired a shot in anger. In summary, it was indeed a great deal of effort for such a small weapon which did not play a decisive role in stopping or even slowing down the Allied advance.

The 8cm *Granatwerfer* 34

The 8cm *schwerer Grantwerfer* 34 (sGrW34), actually 81mm, was the standard medium calibre mortar of the German Army and was based on the fairly conventional Stokes and later Brandt patterns, but with some modifications. A total of six were issued to the machine gun company of an infantry battalion and fifty-four distributed between an entire infantry division in June 1941. The three component parts of the weapon, barrel, baseplate and bipod, could be carried by members of the crew. Alternatively, the parts could be transported on a horse-drawn cart, along with twenty-one rounds of ammunition ready to use. In such cases, one member of the *Granatwerfergruppe* of two weapons was termed the *Pferdführer* or horse-leader, and it was his responsibility to make sure the weapons were ready when needed. On reaching the destination each of the components of the mortar would be unloaded and

8cm mortar projectiles		
Type	*Colour*	*Weight*
HE Wgr34	Maroon body	7.72lbs
FS smoke Wgr34Nb. & 38Nb.	Maroon body, white Nb	7.72lbs
Bounding HE Wgr39Umg.	Green body	7.72lbs
Marker Wgr38Deut. (blue smoke)	Field grey body	7.72lbs
Note: Wgr is a mortar bomb or quite literally in German 'thrown shell'. In German usage, *Granate* (grenade) refers to a high explosive shell.		

carried to the firing position. When going into combat in support of infantry, the weapon would be transported by armoured half-track carriers such as the SdKfz 251/2 or the SdKfz 250/7, which were designated specifically for that role in *Panzergrenadier* units. As with any such combat deployment, and despite regulations specifying load capacity, mobile units would have carried as much ammunition as possible aboard the vehicles to ensure the mortar could keep firing for as long as possible. The sGrW34 could be traversed between 9 and 15 degrees on its bipod according to the angle of elevation, and fired HE bombs weighing 7.5lbs out to a maximum range of 2,625 yards, with a rate of fire in the order of fifteen rounds per minute. When it was fired on soft ground, the baseplate was very often not set properly. Instead, the practice was for a couple of members from the crew to hold the bipod legs during the first few firings until it became firmly embedded. The mortar was provided with two types of HE bomb, the *Wurfgranaten* 38 and the *Wurfgranaten* 39, both of which were designed as 'bouncing bombs' to produce airbursts without having to set a time-fuse delay. The heads of these bombs were filled with a charge of smokeless powder separate to the main burster charge in the body of the actual round. When the bomb landed nose-first, the impact ignited the charge in the nose of the bomb and this threw or 'bounced' it into the air to a height of some 15 to 20ft, at which point it detonated, throwing

Figure 50. Sight unit fitted to German GrW34 8cm mortar. (*SASC*)

shards of steel in all directions. It was highly effective against troops in the open advancing across firm ground, but when the bombs landed on soft surfaces the results were often less than satisfactory. The British Army tried to develop a similar bomb for its 3in mortar but the idea was soon dropped because of the poor results when fired on to soft ground.

In its original version the sGrW34 had a combat weight of 125lbs, which was considered too heavy for airborne troops, so it was decided to develop a lightweight version for these specialised units. This was the short-barrelled *kurzer* 8cm GrW42, nicknamed *Stummelwerfer* (Stump-Thrower). It was usually issued to *Fallschirmjäger* and *Gebrigsjäger*, but in an emergency could be issued to regular infantry battalions when there was no other alternative available. In such cases it was used in conjunction with standard GrW34 in support units. The modifications produced a weapon with a barrel length of only 29.4in, as opposed to the original barrel of 45in, with a weight in action of only 62lbs. As expected, the modifications meant the range dropped, to 1,200 yards, less than half its original range, despite the fact that it fired the standard ammunition. However, the loss of range was acceptable because the mortar's light weight meant it was easier to handle than the original weapon while maintaining the same rate of fire. A scaled-up version of the 8cm, the 10cm (actually 10.5cm) *Nebelwerfer* 35 (NbW35), initially served with *Nebeltruppen* (Smoke Troops). It was replaced by *Nebelwerfer* rocket launchers in 1942, but remained in use as a conventional mortar, firing HE and smoke rounds.

All mortars experience duds, when the bomb fails to fire and leave the barrel. The crew have to deal with such misfires by unloading the weapon and placing the bomb at a safe distance. In the case of the GrW34, the barrels of both the standard version and the shorter *kurzer* designs had a safety catch fitted into the ball-shaped section at the breech end of the barrel. This was a feature designed to allow the crew to deal with a dud bomb in the appropriate manner. According to the manual, in the event of such an occurrence the crew had to wait for one minute. Naturally in combat this was not acceptable and the crew would deal with the problem as quickly as possible. First, the safety catch, marked with the letter 'S' to indicate *sicher* or safe, had to be applied. Next, the barrel clamp on the bipod had to be loosened so the barrel could be rotated to unlock it from the baseplate. A crew member then lifted the breech end to allow the bomb to slide out and be caught by another crew member. The weapon was then returned to its firing condition, sights reset and operations resumed.

The machine gun company organic to a rifle battalion had a platoon of six 8cm mortars. Each mortar troop consisted of six men. In late 1943, the new 'type 1944' grenadier division reorganisation saw two 8cm mortars assigned to grenadier companies. In October 1942, some infantry regiments and rifle battalions and companies were given the designation 'grenadier'. By late 1944, grenadier battalions, on the ratio of two battalions per regiment, were assigned a heavy company – rather than the old machine gun company – with a platoon of six 8cm or 12cm mortars. Ammunition for the sGrW34 mortar was usually carried in pressed-steel containers, each holding four complete rounds of ammunition. The three-man crew carried twenty-four rounds of ammunition ready to use, with additional supplies being carried either in support transport or by other members of the company, who could carry containers into the mortar positions ready for immediate use. This was a widespread practice in most armies during the war and, in fact, is still used today.

The *schwere* 12cm *Granatwerfer* 42

In the opening phases of Operation Barbarossa in June 1941, the German *blitzkrieg* into the Soviet Union not only led to huge numbers of Soviet troops being taken prisoner, but also vast stocks of abandoned weapons being seized. Among the captured stockpiles the Germans discovered large numbers of the 120mm calibre PH38 mortar. Never ones to overlook an opportunity, they seized on the chance to absorb the 120mm Soviet mortars, along with ample stocks of ammunition, which had formerly been used by Soviet artillery units, and took them into service as the GrW378(r). Once crews had been trained in the use of these weapons, they were deployed to serve with heavy companies of infantry regiments in place of the 7.5cm and 15cm infantry guns which had previously been equipping what had been the regimental infantry gun company. What made this case different, however, was the fact the Germans were so impressed with the Soviet mortar that they even went to the length of making direct copies of the weapon and standardised it as the *schwere* 12cm *Granatwerfer* 42 (GrW42), the first units of which were issued in late 1942. The German copy was not a straight-forward reproduction, but a modified version which incorporated several small but significant features. For example, the Germans increased the maximum elevation of the barrel from

12cm mortar projectiles		
Type	*Colour*	*Weight*
HE	Brown	38.84lbs
HE (Soviet)	OD	35.3lbs

the original 80 to 85 degrees and allowed for greater traverse on the bipod, which was set to between 8 and 16 degrees, depending on the angle of elevation set on the barrel. The mortar was fitted with road wheels for easier transportation and the Germans modified the design of this feature to around 22in width, giving a more stable wheel base than the slightly narrower Soviet design of 20in. Other modifications included a heavier baseplate and a sturdier bipod, all of which increased the weight by 100lbs to produce a weapon that weighed 1,234lbs in the transport mode – heavier than the 5cm PaK38 anti-tank gun. Despite this, the new mortar was well liked by the crews, who were satisfied with its performance and handling capabilities.

The weapon was fairly conventional in its layout, and in combat without the transport wheels, the weight of the equipment dropped to 628lbs, which was really the upper bracket for such a weapon in use with front line troops. The barrel was 73.5in in length and weighed 231lbs, with the bipod and baseplate weighing 154lbs and 243lbs respectively. It fired 34.83lb HE bombs in the standard drop method, with a muzzle-loading action. A crew could achieve a firing rate of fifteen rounds per minute, with a maximum range of 6,615 yards.

The transport unit was constructed from steel tubing with pressed steel wheels fitted with pneumatic tyres, and had a towing eye to permit the weapon to be moved by any vehicle equipped with a tow hook. The forward end of the transporter incorporated a circular clamp for the barrel. The rear end was 'u-shaped' and slotted into brackets on the baseplate to provide a secure framework when being towed. Despite its cumbersome size and weight, the *Granatwerfer* 42 could be moved into and out of action with relative ease. When the crew wanted to move out they simply lifted the baseplate from the firing position, swung the barrel down to a vertical position and moved the bipod round so that it lay folded along the barrel. They then pushed the transporter into position and locked the mortar in place with the securing couplings on the baseplate and barrel clamp. The weapon was now ready for towing to its next location. To bring the weapon into action, the crew simply reversed the process and started firing once ready. By 1943, the 12cm had begun to replace the six 7.5cm and two 15cm infantry guns organic to the infantry regiment in many units. In this case the infantry gun company was re-designated a heavy company.

The Italian Army

The Italian Army conducted five major campaigns during the war, if one includes the fighting in mainland Italy by those Fascist forces remaining loyal to Mussolini after he was deposed in 1943. Some Italian troops had already gained an amount of combat experience prior to the outbreak of war, during the Spanish Civil War and the campaign in Ethiopia in North Africa. The aborted attempt to join Germany in attacking France in 1940 was an embarrassment and North Africa was a long, drawn-out campaign which also ended in disaster. Italy's incursion against Greece, which required German intervention, proved how inadequately prepared the Italian Army was for war. Mussolini's decision to support Hitler's invasion of Soviet Russia resulted in heavy losses, again due to poor equipment and weaponry unsuited to such extremes of weather. Even so, the Italians did as well as any soldier could, given his standard of equipment.

Two main designs of mortar stood out from the rest of the Italian arsenal, and these entered service as the standard weapons for the army in 1935. There were other infantry types in service, but these are usually overlooked. A very heavy type, the 240mm calibre *Bombarda*, fired a bomb weighing 144lbs, but this placed it more in the category of artillery weapons than infantry mortars. The Italians had two types of mortar in the 50mm calibre range – the CEMSA *Modello* 1939 and the Cemsa *Modello* 1940, which weighed 36.37lb and 2lb in action, respectively. The *Modello* 1939 could fire a HE bomb weighing 1.9lbs out to a range of over 700 yards, whilst the *Modello* 1940 fired a HE bomb weighing 1.6lbs out to a range of almost 900 yards. The slightly larger CEMSA 63 had a calibre of 63.5mm with a barrel fixed at a firing angle of 45 degrees. It could fire a HE bomb weighing 4.4lbs out to a range of over 1,500 yards. The 120mm calibre CEMSA had a weight in action of 1,830lbs and fired a HE bomb weighing 37.4lbs. Some sources claim this was a version of the Spanish-built 'Franco', which had a range of almost 7,000 yards, but although apparently trialled by the Italian Army it did not enter service.

The Brixia 45mm Model 35 (*mortáio da 45/5 d'assalto modello 35* [m a 45])

The 45mm calibre Brixia Model 35 was for use at platoon level, and if the German 5cm *Granatwerfer* 36 was considered 'over-engineered', then the Italian Brixia Model 35 was over-complicated for use in front line units. For example, such was the extraordinary firing action of this weapon that it had to be mounted on a special tripod when other mortars of comparable calibre were being aimed and supported by the firer's hand. The weapon weighed 34lbs in action and was carried by one man as a single load on a special backpack. In an emergency the man could deploy the weapon, still attached to the carrying pack, and bring it into action, but the action was so extremely complicated as to make it almost impractical. The barrel of the weapon was 10.2in in length

45mm mortar ammunition		
Type	*Colour*	*Weight*
HE	Yellow body, red fins	1.025lbs
Practice	Yellow body, yellow fins	1.025lbs
Instructional	Yellow body, aluminium fins	1.025lbs

and could be elevated to a maximum of more than 85 degrees. The two-man crew could fire twenty-five rounds per minute out to a maximum range of 585 yards, but the light weight of the bomb, only 1lb – of which just over one-eighth (2.5oz) was the burster charge – caused little damage against a protected target. It could cause some damage against unarmoured vehicles such as light trucks caught in the open, and against troops, especially on hard, stony surfaces, the radius of the blast would have caused wounds and even fatalities. The bombs themselves were too light to incorporate their own propelling charges, and special propellant cartridges had to be loaded separately into the firing chamber from a magazine prior to each firing. The bombs were streamlined with the usual fin-shape of mortar bombs, but lacked the power to inflict any great damage. The fuse was an unusual wind-vane-armed design which was extremely complicated and this coupled with the mortar's unconventional action, led to it being dropped from service shortly after the disastrous Libyan campaign. Even so, it remained in limited service and was used in all campaigns fought by the Italian Army, although even as early as 1942, units were being equipped without the Model 35. Some examples were used by German troops serving with the *Afrika Korps* who designated it the 4.5cm *Granatwerfer* 176(i).

The Brixia Model 35 was breech-loaded, an action not used in any other light mortar design, and the barrel was constructed from two concentric tubes, one being the actual barrel with an outer tube forming the breech cover. Both tubes were fitted with a slot, which had to be aligned before a bomb could be inserted into the breech. It was the job of the loader to insert the bombs into the breech through the opening, after which the firer pushed a lever forward to close it, an action which also loaded a propelling cartridge into the firing chamber from a magazine located in the breech assembly. The firer operated a trigger mechanism to fire the cartridge which, in turn, projected the bomb. After firing, the operator realigned the apertures ready for reloading. The barrel could be depressed to 10 degrees to allow almost direct firing, and to elevate the barrel to greater angles a hand-wheel was fitted which engaged a toothed arc on the barrel. Very simple sights were also fitted. To adjust the range further a gas port or opening could be set to vent out some of the pro-pelling gases, thereby shortening the range. For traversing, the weapon was simply picked up and realigned. As weapons go, this was unarguably one of the worst designs ever to enter service, especially considering the limited range and power of the ammunition fired. Infantry battalions had a weapons company which included two platoons of nine 45mm mortars each. Three mortars were assigned to each of the forty-five-man platoons' three fourteen-man sections. A 45mm was crewed by three men, attached to rifle companies and platoons as necessary.

The 81mm Model 1935 (*mortáio da* 81/14 *modello* 35 [m 81])

Although much of the equipment and weaponry used by the Italian Army appeared old-fashioned and, in some cases, obsolete, the 81mm calibre Model 1935 mortar was an exception: a very fine weapon that operated in the con-ventional mode. Essentially it was the same as the American 81mm M1 mortar, and likewise based on the Brandt design. Without any moving part, there was nothing to over-complicate the design or to go wrong. The Model 1935 was slightly smaller, and therefore lighter, than the American mortar, but it had a greater range by over 1,100 yards with its light HE bomb, which weighed slightly more than the American 81mm bomb. The barrel was 45.3in in length and weighed 47lbs, with the bipod weighing 39.7lbs and the base-plate 44.1lbs, to give a weight in action of 130.8lbs, compared to 136lbs for the American M1. The Model 1935 could fire the light HE bomb, weighing 7.2lbs, out to a maximum range of 4,430 yards, and the heavier 15.13lbs HE bomb could reach ranges of 1,640 yards. The crew could fire eighteen rounds per minute, which meant that a support company could saturate a target area most effectively, screen movements of their own troops with smoke bombs and light up areas of the front at night with the parachute illuminating flares to reveal targets.

This weapon saw service in all theatres of combat in which the Italian Army was deployed, including North Africa, Greece, Sicily, Italy itself and on the

81mm mortar ammunition		
Type	*Colour*	*Weight*
Light HE	yellow body, red band	7.2lbs
Heavy HE	yellow body, red band	15.14lbs

Eastern Front. Despite this, the Italians believed that the 81mm Model 1935 lacked sufficient range for desert warfare and even began to withdraw them from some units in 1943. Commonwealth forces thought differently and valued captured Model 1935s because they had a longer range than their own 3in mortars which fired out to 2,800 yards. The British and Commonwealth troops dreaded mortar barrages, during which all they could do was take shelter in their slit trenches. Veterans recall hearing the faint 'popping' sounds as the mortars began firing off in the distance, followed a few seconds later by a noise which sounded like 'an express train' and an earth-shattering explosion. The effects of the blasts were multiplied by stones and rock shards being thrown out along with shell splinters. The Indian troops of the Rajputna Regiment in the fighting in the Dolgorodoc heights, during the Keren campaign in Ethiopia (5 February to 1 April 1941), experienced the firepower of these weapons and learned how hard the Italian soldiers could fight. The Indian troops had to take the prepared defences by hand-to-hand fighting with grenades and bayonets.

A standard infantry division of the Italian Army had a 529-man mortar battalion of three companies, each with nine 81mm mortars, to give the unit an allocation of twenty-seven weapons. From 1942, these battalions were often armed with only eighteen mortars, six per company. During the early stages of the war, infantry regiments were structured to include nine mortars, with some 199 troops to operate them, but later some units had only six mortars and the troop level was also reduced. A typical company had three platoons, each with three sections equipped with one mortar. From 1942, a number of regiments had no 81mm mortars and instead had to rely on weapons to be attached from the divisional mortar battalion to provide fire support. When Italy surrendered on 3 September 1943, the Germans disarmed those Italian troops not still loyal to the deposed Mussolini, who was established as the puppet leader of the equally puppet Salo Republic. Those Fascist troops who remained loyal and continued to fight as Germany's ally used the Model 1935 mortar until the end of the war.

The Japanese Army

Imperial Japan began expanding its influence by making military movements against China as early as 1931, when Japanese troops occupied Manchuria. Three years later, Japanese troops seized the province of Jehol and in 1937 the situation between the two countries had developed into all-out war. Japan had been allied to Britain and France in the First World War but had chosen to go in its own direction after leaving the League of Nations in 1933. All of

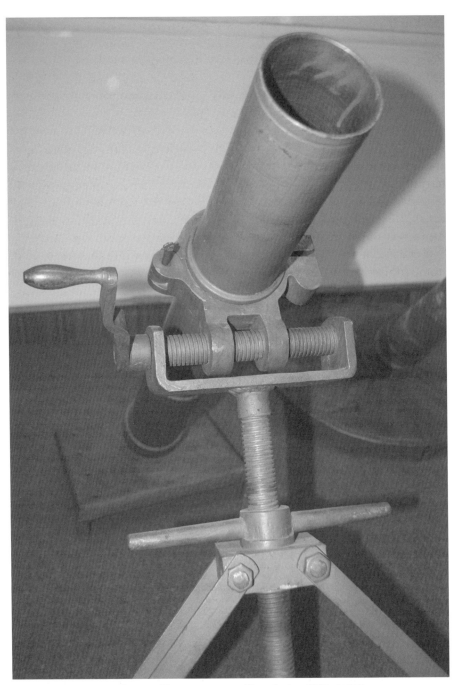

Figure 51. Experimental type of Japanese mortar Type 3 *circa* 1929. (*SASC*)

this had happened in the eighty years since Japan had been forced to open the country's doors to international trade at gunpoint by Commodore Matthew Perry of the US Navy. During that time the country's military power had grown, with the Imperial Japanese Navy developing from being a 'white-water' (coastal) navy to a 'bluewater' (deep sea) navy good enough defeat the Russian fleet in battle at Port Arthur in 1904.

Japan entered the war by attacking the American fleet at Pearl Harbor. They rapidly expanded across the Pacific to set up a defensive perimeter, securing an area covering some 2,000 miles by 2,500 miles. This took in the occupied territories of mainland China, the Malay Peninsula and the many islands scattered across the ocean. Garrisons were deployed to these islands in order to defend the newly-won conquests. Whilst an infantry division of the Japanese Army was organised along lines similar to their European and American equivalents, an army could vary in size from 50,000 to 150,000 troops. Units from within one particular division could often be taken and reallocated to another for a specific role or operation. An infantry battalion had 1,035 men, with three rifle companies of 195 men divided into three platoons with four sections and a separate platoon with mortars. The size of a Japanese infantry division between 1941 and 1945 hardly altered, but the manpower levels would drop as the Allies gained strength, recaptured territory and cut off the supply lines of raw materials and fuel. The size of the army would peak at around 5 million men in 140 divisions which fought against the British, Indian and US forces from Burma and Malaya to the islands across the Pacific Ocean. The cost would be high for the length of time they were engaged in fighting, and Japan lost around 1.7 million men in the field.

In order to secure Japan's newly-won territories, the Japanese established a defensive perimeter around its limits and parcelled off units to the various islands, some of which had little or no military value either strategically or tactically. The size of these garrisons could also vary, with some comparable in size to brigades (between 3,000 and 5,000 troops) and others the equivalent of an army, such as the force of 120,000 troops on Okinawa. The garrison on Tarawa numbered only 2,619 troops, with a construction force of 2,200 workers to build the defences. Only seventeen Japanese surrendered after the three-day battle to capture Tarawa. On Tinian the garrison was 8,800 infantry and marines, which was equivalent to two brigades. The Japanese Army did not have an equivalent organisation to a corps (usually around 30,000 to 50,000 troops), but some of the island garrisons, such as Saipan and Guadalcanal, which had 31,000 and 36,200 troops respectively, could have been classified as such levels. Naturally, problems arose over supplying these disparate island garrisons, which were dotted across the vastness of the Pacific Ocean. The Japanese used some of the islands as airbases to operate against Allied naval forces. As a counter-measure, the Americans developed a unique strategy to tackle the difficult problem of dealing with so many potential targets. The strategy was called 'Island Hopping' and required amphibious landings against each of those islands being attacked.

Japan had been re-equipping its forces since the 1920s – far longer than most European countries – and, having no border, it did not have to concern itself with alarming neighbouring states. The Western powers of Britain, France and America did not perceive any threat and saw the modernisation and re-armament programme as progressive. By late 1941, and the attack on Pearl Harbor, the Japanese had many modern aircraft in service, the navy had expanded and the army had new weapons, including several designs of mortar, either in service or on the point of entering service. These replaced ageing weapons which had been declared obsolete, such as the 5cm calibre Type 98 which fired a box-like demolition charge of 14.1lbs containing 7lbs of shimose, a compound based on picric-acid. During the war, the Japanese Army used up to eleven main types of mortar ranging in calibre size from the light 5cm designs for platoon level use through to the upper-limit 9cm calibre mortars which were in company support units, and the very heavy 150mm calibre Type 93 and Type 97. In between there was the Type 11, introduced into service in 1922, which was a 70mm calibre weapon. This weighed almost 134lbs and was capable of firing HE bombs weighing 4.6lbs out to a range of 1,700 yards, which was acceptable for fire support. Probably as few as 234 of this type were ever produced. In the first months of the war, the Japanese came to be considered virtually unstoppable because of their ability to cross vast tracts of jungle wilderness at speeds which took the Allies by surprise. The Japanese, as we now know, were not master tacticians of jungle warfare; it was simply they did not fear the jungle as a theatre and they used this to their advantage. The myth was not to be broken until the uncon- ventional tactics of General Orde Wingate established the British Chindits as a fighting force which continued on the offensive in all weathers.

Indeed, the British commander in Burma, Lieutenant-General William Slim, later Field Marshal 1st Viscount Slim, wrote of the conditions: 'To our men, British and Indian, the jungle was a strange, fearsome place; moving and fighting in it was a nightmare. We were too ready to classify jungle as "impenetrable", as indeed it was with our motor transport, bulky supplies and inexperience. To us it appeared only as an obstacle to movement and vision; to the Japanese it was the welcome means of concealed manoeuvre and surprise. The Japanese used formations specially trained and equipped for a country of jungles and rivers, while we used troops whose training and equipment, as far as they had been completed, were for the open desert. The Japanese reaped the deserved reward for their foresight and thorough preparation; we paid the penalty for our lack of both.'

It was in the harsh terrain and dense foliage of the jungle in the Far East that the mortar rose to its full significance as a weapon offering infantry fire support in the absence of conventional artillery. Tanks and artillery could not be moved freely in such conditions, and so it was left to the infantry to use whatever means necessary to engage the opposition, and this is where mortars came fully to live up to the designation the 'artillery of the infantry'. In areas such as Burma, Borneo and Malaya there were no readily-defined front lines, as in North Africa and later Europe. It was a war of infantrymen, who in the case of the Chindits were supplied by air drops delivering food, medical aid and ammunition that allowed the troops to operate for longer periods than usual. In the jungle, the Japanese soldier deployed his small arms and mortars to fight the Allies. It has been rightly claimed that mortars were a vital factor in jungle fighting, which at times was conducted at very close quarters. The weapons were used to provide fire support, as they could be carried into areas where artillery could not easily be deployed, if at all.

Like other armies during the Second World War, the Japanese deployed mortars around infantry units in their own unique manner. Mortars were used by all front line units of the Japanese Special Naval Landing Force, along with any naval units which were deployed to serve in land operations. Even those specialised forces such as amphibious and airborne troops were equipped with 5cm and 8.1cm mortars. A strengthened triangular infantry division had 251 5cm grenade dischargers and no conventional infantry mortars, although a mortar battalion as discussed below could be attached. The Japanese initially relied on 70mm battalion guns and 75mm regimental guns for fire support. Later in the war, these were often replaced by 8cm mortars as they were easier to operate and could be manufactured faster at a lower cost. Mortars were better suited for the close-range defence tactics they adopted in the Pacific.

An Independent Mixed Brigade, with a troop strength of almost 5,600 men, had two companies tallied off to serve as support units, and these would be equipped with 81mm mortars or the larger Type 94 mortar of 9cm calibre. The Japanese also established mortar battalions, known as *Hakugeki Daitai*, which were equipped with thirty-six mortars of 8cm calibre. This was a self-contained support unit with its own integral headquarters with signallers, observers and transport, which, depending on terrain and availability, could either be horse-drawn or motorised. Each such battalion had its own ammunition train, and was formed into three companies, each with twelve mortars. A typical infantry battalion of the Japanese Army had a troop strength of 580 men, of which 114 men were formed to make machine gun and infantry gun companies equipped with two 3.7cm anti-tank guns and two 7cm howitzers. In the absence of the howitzers, mortars could be used, at which time the unit became termed a *Kyuho* group. Until 1943, battalion structure mortar units, known as *Hakugeki*, were a standard part of any infantry battalion. But after that date, when the war started to go very badly for the Japanese, such mortar units could be detailed to serve with any infantry unit, even those with less than company strength.

The 50mm Type 89 (1929)

The Type 89 (1929), also known as the *hachiku shiki tekisanto* (89 model heavy grenade discharger), was a 50mm light mortar and one of the mainstay mortars of the Japanese Army. Termed a 'grenade discharger', due to the fact that it fired an adapted version of the standard infantry Type 91 hand grenade, it could also fire a dedicated mortar bomb. This was the weapon which caused the Allies a great deal of anguish when they first encountered it, not on the battlefield, but when they came to test-fire captured examples in trials to assess its capability. It appears that a mistake was made during the translation of captured field manuals and the Type 89 was called the 'knee mortar'. In fact it should have been the 'leg mortar' because it was carried strapped to a man's leg. This error, coupled to the fact that it was fitted with a

5cm grenade discharger ammunition		
Type	*Colour*	*Weight*
HE Model 89	Black body, red and yellow bands	1.75lbs
Incendiary	Natural steel body	1.25lbs
WP smoke Models 11 and 95	Brass body	
Coloured flare Model 10		
Coloured smoke Model 10		
Practice Model 91 and 94		
Dummy Model 94		
Frag grenade Models 10 and 91	Black body, red top	1.175lbs
WP grenade Models 10 and 91	Brass body	1.1lbs

convenient curve-shaped spade at the end of barrel, which slotted perfectly over a man's knee, led some servicemen to believe the firing position involved placing the weapon on a bent knee. Experienced soldiers should have known better, yet some did fire the Type 89 in this manner. The action of the recoil resulted in severe broken legs to those who misguidedly attempted this foolish method of operation.

The Type 89 was unusual because its 10in long barrel was rifled with eight grooves of right-hand twist. Apart from that, the weapon had the appearance of being a standard light mortar intended for use at platoon level. It weighed 10.1lbs in action and was fitted with the trip-style firing mechanism associated with other such light mortars. Being hand-held when fired, it could be used at a range of angles, all judged by the firer's line of sight, but usually above 45 degrees. It had a maximum range of 700 yards and could fire up to twenty-five rounds per minute. The ammunition used for the Type 89 was either the standard mortar-type high explosive shell, weighing 1.75lbs, or the specially modified Type 91 hand grenade, which weighed 1.5lbs. There was also a wide variety of coloured smoke bombs – the Japanese used some fifteen different types, known as 'dragons' to the Japanese troops – and also flare signals which could be used to direct heavier mortars or artillery onto targets. It was a surprisingly simple weapon despite being fitted with a rifled barrel and, as far as the front line troops were concerned, it was far and away the most popular mortar for use at platoon level. The Japanese Army had few opportunities to use its parachute troops but one of its first operational deployments was when airborne units landed in Palembang Sumatra, in February 1942. This was followed up quickly when another paratrooper force was dropped on Leyte. The paratroopers were armed with personal weapons and Model 89 (1929) 50mm grenades which could be fired from the Type 89 mortar, which they also carried. They also carried a range of other equipment in bags suspended from a harness, and these contained further rounds for the mortar and hand grenades.

Range control for this weapon was achieved by simply adjusting the support rod, which carried the firing pin, by screwing it further into the barrel, thereby creating a variable chamber space into which the gases of the propelling charge could expand on firing. Thus, when loaded, the bomb only dropped as far as the adjusted support rod would permit. This meant that a large space had to be filled by propelling gas, which led to a drop in pressure, thereby causing a reduction in range. The reverse was true when it was a smaller chamber and the range was increased. The Type 89 mortar fired a standard bomb designed specifically for the weapon. This was a self-contained round of ammunition, featuring a fuse, propelling charge in the base, in addition to which it was fitted with a copper driving band to engage the rifling of the barrel. The mortar was muzzle-loaded in the standard way and was fired using a tripping action lever. The driving band around the bomb was of small enough diameter not to touch the rifling, permitting ease of loading. On firing, some of the propelling gases were vented from a series of holes around the body of the bomb, which forced the driving band to engage the rifling. This reduced the amount of windage, the loss of propulsive force, due to poor obturation between the body of the bomb and the internal surface of the barrel. If no ammunition of this type was available, then standard infantry hand grenades of the Type 91 design could be utilised via a conversion kit.

The conversion kit to ready the hand grenade for use with the Type 89 discharger was comprised of a finned tail section with primer charge and propelling charge which was screwed into the base of the grenade. To make it suitable for mortars, the Type 91 grenade was fitted with a fuse which had a burn time of 7.5 seconds as opposed to the more usual 4.5 seconds of other Japanese hand grenades, such as the Type 97 and 'Stick-type'. The modification lacked range due to the poor gas seal around the improvised projectile, but what it lacked in range it more than made up for in lethality. The Type 91 grenade contained only 2oz of TNT as a burster charge, but it was enough to hurl large fragments of the cast iron body and inflict serious injuries in the close-quarter fighting of the jungle.

The Japanese retained the earlier 50mm Type 10 (1921) grenade discharger, which was a simpler and much lighter weapon, commonly called the *jutcki*, also known as the Type 10 *jnenu nenshiki tekidanto* (10th year model grenade discharger). This smoothbore weapon – introduced in 1921 – could not fire the mortar-type rounds used by the Model 89 with its rifled barrel, and was restricted to firing modified hand grenades and smoke and flare signals. The Type 10 weighed 5.25lbs in action and had a range of around 175 yards, which was much further than a man could throw a hand grenade. The 50mm calibre Type 99 platoon weapon weighed 10.25lbs in action and could fire a HE bomb weighing 1.75lbs out to ranges over 700 yards, making it ideal for use in the jungle. These light calibre mortars were in effect grenade dischargers and were allocated on the basis of one to each of a rifle platoon's three light machine gun sections (rifle squads) and two in the platoon's grenade discharger section – essentially a rifle section without a machine gun. Although light, they were effective in providing fire support during battles among the dense foliage of the jungle, where firefights could take place at very short ranges.

The 81mm Model 97 and Model 99

The Type 97 (1929) was designated the *shiki kyokusha hoheiho* (97 model high-angle infantry gun) while the shorter Type 99 (1939) was the *kyukyu shiki shohakugekih* (99 model small trench mortar). This weighed only 53lbs in use, comprising the bipod, which weighed 17lbs, and the barrel and baseplate, each of which weighed 18lbs, and could fire a high explosive bomb weighing 7lbs out to ranges of 2,000 yards. From this design the Japanese Army developed a modified weapon with an increased range. This entered service in 1937 as the Model 97 mortar with the almost universal calibre of 81mm, but which the Japanese rounded down to 8cm, in the same way their German allies were inclined to do. The new Model 97 was based on the Stokes-Brandt design, and in all respects was virtually identical to those weapons of comparable calibre in service with the American and Italian armies. Around 600 were produced. Rather confusingly, the Japanese also referred to this weapon as the 'Model 37 High Angle Infantry Gun' despite the fact that as a mortar it fired bombs in the traditionally high elevations of angle associated with such weapons, which is to say between 45 and 90 degrees, while a field gun fires below 45 degrees of elevation. The Model 97 weighed over 145lbs in action and the barrel was 49.5in in length. The weight and size of the weapon did not make it popular with the troops who had to carry and use it in very harsh terrain.

A naval version of the Type 97, with the same length barrel, was developed for the Imperial Japanese Navy and Marine troops. This was known as the Type 3 and is believed to have been an experimental design. The barrel of the Type 3 mortar could be adjusted to fire at angles between 45 and 85 degrees, and like the standard Type 97 design, it could fire a light HE bomb weighing 7lbs out to a range of 3,300 yards and a heavy HE bomb weighing 14.3lbs to ranges of over 1,300 yards. It weighed almost 165lbs in action, 20lbs heavier than the army version, with no perceivable gain in range, and it was probably as unpopular with naval troops and marines as the army version. In 1939, the Japanese Army began taking into service a modified design of the Type 97, known as the Model 99, which could be fired at the rate of fifteen rounds per

8cm mortar ammunition		
Type	*Colour*	*Weight*
Light HE	black body, red, yellow, white bands	7.2lbs
Heavy HE	black body, red, yellow, white bands	14.3lbs

Figure 52. Japanese 81mm Type 99 mortar. (*SASC*)

minute. The weapon was fitted with a much shorter barrel, around 25.25in, which also reduced its weight in action to 52.25lbs, comprising the barrel which weighed 17.5lbs, the bipod a further 16.5lbs and the baseplate weighed 18lbs, which was much lighter than the American M1 or Italian Model 35, which weighed 136lbs and 129lbs respectively for the same calibre. It is understood the Type 99 was developed for Japanese airborne troops who needed a weapon like this with the reduction in weight and size. It later proved ideal for use by the army in the jungle.

The Model 99 mortar was an unusual design because it was fitted with a selective drop-type firing mechanism which included the fixed firing pin method, a standard feature on all 81mm calibre mortars, and a tripping lever action of the lighter calibre weapons. Fitting the secondary type of firing mechanism was probably because of its intended use with airborne forces. Should one form of firing mechanism become damaged during the parachute drop, there would always be a secondary means of firing the weapon. The fixed firing pin used for the standard 'drop' method of firing could be retracted into a special housing until its base rested on a conical shaft which passed across the base cap and protruded from the breech end at a right angle to the axis of the barrel. In this mode it had to be operated by the firer striking it very hard with a hammer, which forced it back into the conical section and in turn struck the base of the bomb to fire it.

These changes made the Model 99 popular with the troops, and apart from having a range some 800 yards lower than that of the original Model 97, the newer version gave a good account of itself in action. The Model 99 could fire all the types of ammunition used by the Model 97, including both the light and heavy explosive bombs. However, no confirmation has yet been found that the Model 99 used the heavy bomb range of the Model 97. This is probably because the modified design could not take the firing pressures of the heavier bomb. Other types of ammunition included white phosphorous smoke bombs, which given the nature of the compound could ignite combustible materials, green signalling flares and parachute smoke signalling and illuminating bombs. Like other weapons of the same calibre, the Model 97 and the Model 99 could fire captured stocks of American 81mm calibre M43 mortar rounds, the latter weapon being able to achieve ranges of 2,500 yards, which was 200 yards greater than with its own standard ammunition.

The 90mm Type 94 and Type 97

The Type 94 (1934) *kyuyon shiki keihakugehiko* (94 model light trench mortar), together with the Type 97, were 90mm calibre weapons whose size placed them between medium support weapons and the heavy calibre types. The two

9cm mortar ammunition		
Type *Colour*		*Weight*
HE black body, red, yellow, white bands		11.5lbs

weapons, while similar, were easily distinguished from one another due to the recoil mechanism fitted to the Type 94. It was this factor which separated the weapons with regard to the difference in weight. The Type 94 entered service in 1935 and weighed 341.5lbs in action, which was considered too heavy for the army in jungle fighting. Probably only around 450 were produced. Around 600 Type 97 mortars were made. They were introduced into service in 1937. This was an altogether much lighter weapon than its Type 94 counterpart, weighing only 233lbs, due to the fact that it did not have the recoil mechanism which accounted for almost a third of the weight of the Type 94. Both weapons fired the same type of ammunition. The disparity of weight led to the Type 94 being considered too heavy and cumbersome for use during the long-range mobile operations conducted in the dense jungles of Burma and Malaya, but it was used in defensive positions on Guadalcanal. The hydro-pneumatic recoil buffer mechanism fitted to the Type 94 was secured to the barrel by means of u-shaped locking pins, and at the top of the bipod were two plungers incorporated indirectly into the buffer system. The maximum length of the recoil stroke was some 5.75in, which gives some indication of the forces generated on firing. Without this recoil mechanism, the force would have been driven the mortar into the ground. The recoil system on the Type 94 would have made it an ideal weapon for mounting on a modified half-track such as the Type 98 Ikegai KO-HI, an artillery prime mover, to provide a useful self-propelled mortar system. The Japanese method of conducting battle did not make allowances for such a system, however, and the garrisons on the small islands did not require such vehicles.

Both mortar types fired a HE bomb weighing 11.5lbs, and whilst the range achieved with the Type 94 was over 4,000 yards, the Type 97 had a maximum range of 4,150 yards, which to the troops meant better fire support. In the end, though, they had to use whichever design of mortar was available to them at the time. Each model had a traverse of 10 degrees from the centreline and fired at angles of elevation between 45 and 70 degrees. Their rate of fire could be in the order of fifteen rounds per minute, which made each design useful support company weapons, especially when firing as a full battery to support an attack. As well as the standard explosive bomb, the Japanese developed an incendiary bomb, which could be used to set fire to combustible material such as dry vegetation. During operations in China, Japanese troops fired these bombs against the flimsy, densely-packed huts in Chinese villages. These bombs contained around forty pellets of an incendiary composition made up of white phosphorous combined with carbon disulphide – a particularly lethal combination. When the bomb landed, the 'burster' charge detonated it, rupturing the casing, the blast scattering the pellets in all directions to ignite soft targets such as ammunition dumps and vehicles. The action also had an anti-personnel effect against troops in defensive positions or moving across open spaces. The Allies used their white phosphorous smoke bombs to set fire to vegetation as a secondary effect, but here the Japanese had developed mortars for the specific purpose of being incendiary bombs.

After the War

From humble beginnings as a stop-gap weapon in the trenches of the First World War, the dependency on the mortar grew to provide fire support. It was used in all campaigns and in all theatres of fighting. Mortars played their part right the way through the six years of the war, giving service to each army in all conditions of weather and terrain. For instant fire support, the mortar was unsurpassed on the battlefields of the Second World War. The infantry and lightly-armed airborne units were glad of the support these weapons provided, and cursed those of the enemy.

In the years immediately after the war, many of the types of weapon remained in service with their countries of origin and were later supplied to overseas armies. Many wartime weapons were used in large-scale conflicts like the Korean War, where the Communist forces used mainly Soviet-built weapons and the United Nations forces used 81mm calibre mortars and in the case of the British Army the 3in mortar. Right through to the 1970s conflicts in the Middle East between Israel and Egypt, Second World War weapons were still in use. As these weapons wore out and became obsolete, armies replaced them with modern types which were more powerful.

The service mortars render to armies continues today as a direct result of developments made during the Second World War. For example, self-propelled mortar vehicles are in service with modern armies around the world, and many types have been used in combat. Infantry mortars still fire the three basic types of ammunition, though technology has led to the development of specialist bombs which can not only be used to engage armoured vehicles, but are fitted with electronic sensors that allow these projectiles to seek out targets. Ranges have increased and today there are some types which can be

Figure 53. Soviet-built M37 82mm mortar used in the Second World War and captured by British troops in Afghanistan in June 2002. (*SASC*)

categorised as 'high power light mortars' with ranges in excess of 5,000 yards, and also long-range towed mortars with ranges of over 6,500 yards. Mortars have been used in hi-tech conflicts such as Iraq and Afghanistan where drones, supersonic aircraft and tanks, along with artillery, dominate the action. Battlefield radar used to track mortar projectiles has developed to such a degree that firing positions can be located with pinpoint accuracy within fractions of seconds, something only made possible by the ground-breaking work which was started in the Second World War.

The historical reference to mortars can be traced back many centuries, but it is only in the 100 years since the First World War, that they have become indispensable for providing fire support to infantry units. They can be air-lifted by helicopters and flown straight into combat situations, and some are so light they can easily be carried and operated by one man. During the war, the mortar was known as the artillery of the infantry. This role continues today and will do so for many years to come. The mortar, which once resembled nothing more than a length of drainpipe, is still very much a front line weapon.

Bibliography

Bailey, R.: Forgotten Voices of D-Day, Ebury Publishing, London 2009.

Barker, A.J.: Ian Allan, Shepperton, Surrey, 1979.

Beevor, Antony: Stalingrad, Penguin Books London, 1999.

Bredin, A.E.C.: Three Assault Landings, Aldershot, 1946.

Chamberlain, Peter and Doyle, Hilary: Encyclopaedia of German Tanks of World War Two, Arms & Armour, London, 1999.

Chamberlain, Peter and Ellis, Chris: Gebirgsjager, Almark Publishing, London, 1973.

Chinnery, Philip D.: March or Die, Airlife Publishing Ltd, Shrewsbury, 1997.

Clark, Lloyd: Arnhem; Jumping the Rhine 1944 and 1945, Headline Publishing Group, London, 2009.

Comparato, Frank E.: Age of the Great Guns, Stackpole Company, Harrisburg, Pennsylvania, 1965.

Crociani, P. and Battistelli, P.P.: Italian Soldier in North Africa 1941–43, Osprey Publishing, Oxford, 2013.

Davies, W.J.K.: Ian Allan, Shepperton, Surrey, 1973.

Edwards, Roger: German Airborne Troops, MacDonald & Janes, London, 1974.

Forty, George: Battle for Monte Cassino, Ian Allan Publishing, Hersham, Surrey, 2004.

Ginns, Michael and Bryans, Peter: German Fortifications in Jersey, Meadowbank, St Lawrence, Jersey, Channel Islands, 1978.

Greentree, David: British Paratrooper versus Fallschirmjager, Osprey Publishing, Oxford, 2013.

Haskew, Michael E.: Small Arms 1914–1945, Amber Books, London, 2012.

Hogg, Ian: Mortars, Crowood Press Ltd, Marlborough, Wiltshire, 2001.

Hogg, Ian: Fortress A History of Military Defence, MacDonald and Jane's, London, 1975.

Hogg, Ian: The British Army in the 20th Century, Ian Allan, Shepperton, Surrey, 1985.

Holmes, D.C.: Fortress Jersey, German Army, self-published, 1978.

Holmes, Richard: Firing Line, Jonathan Cape Ltd, London, 1985.

Howard, Gary: America's Finest II, GHQ Militaria, Tonbridge, Kent, 2005.

Jackson, Robert: Dunkirk The British Evacuation 1940, Cassell, London, 2002.

Keegan, John and Holmes, Richard: Soldiers A History of Men in Battle, Sphere Books Ltd, London, 1987.

Kershaw, Robert J.: D-Day Piercing the Atlantic Wall, Ian Allan, Shepperton, Surrey, 1993.

Lee, David: Beachhead Assault, Greenhill Books, London, 2004.

Linklater, Eric: The Defence of Calais, His Majesty's Stationery Offices, London, 1941.

Lunde, Henrik O.: Finland's War of Choice, Casemate Publishing, Oxford, 2013.

Norris, John and Calow, Robert (illustrator): Infantry Mortars of World War II, Osprey, Oxford, 2002.

Norris, John: World War II Trucks and Tanks, Spellmount, Stroud, Gloucestershire, 2012.

Reid, William: The Lore of Arms, Purnell Book Services Ltd, Abingdon, Oxon, 1976.

Rossiter, Mike: We Fought at Arnhem, Transworld Publishers, London, 2012.

Shaw, Frank and Shaw, Joan: We Remember Dunkirk, Ebury Press, London, 2013.

Smith, Claude: History of the Glider Pilot Regiment, Pen & Sword Books, Barnsley, South Yorkshire, 2014.

Swaab, Jack: Field of Fire, Sutton Publishing Ltd, Stroud, Gloucestershire, 2005.

Tootal, Stuart: The Manner of Men, John Murray (Publishers) London, 2013.

Whiting, Charles: '44 In Combat on the Western Front from Normandy to the Ardennes, Century Publishing, 1984.

Womer, Jack and Devito, Stephen C.: Fighting With the Filthy Thirteen, Casemate Publishers, Oxford, 2012.

Young, Edward M.: Meiktila 1945, Osprey Publishing, Botley, Oxford, 2004.

Zaloga, Steven J.: M7 Priest 105mm Howitzer Motor Carriage, Osprey, Botley, Oxford, 2013.

Index